Yellowcake and Crocodiles

Yellowcake and Crocodiles

Town Planning, Government and Society
in Northern Australia

John P. Lea
Department of Town and Country Planning,
University of Sydney

Robert B. Zehner
School of Town Planning,
University of New South Wales

ALLEN AND UNWIN
Sydney London Boston
Northern Territory University Planning Authority

For
Dee and Ruth
Mary-Anne, Alexandra, David and Eric

First published in 1986
Allen & Unwin Australia Pty Ltd
8 Napier Street, North Sydney, NSW 2060 Australia

Allen & Unwin New Zealand Limited
60 Cambridge Terrace, Wellington, New Zealand

George Allen & Unwin (Publishers) Ltd
18 Park Lane, Hemel Hempstead, Herts HP2 4TE England

Allen & Unwin Inc.
8 Winchester Place, Winchester, Mass 01890 USA

National Library of Australia
Cataloguing-in-Publication entry

Lea, John P. 1940–
    Yellowcake and crocodiles.

    Bibliography.
    Includes index.
    ISBN 0 86861 875 6.
    ISBN 0 86861 843 8 (pbk.).

1. Sociology, Urban — Northern Territory — Jabiru. 2. City planning —
Northern Territory — Jabiru. 3. Uranium mines and mining — Social aspects —
Northern Territory — Jabiru. I. Zehner, Robert B. II. Northern Territory.
University Planning Authority. III. Title.

307.7'66'0994295

Set in 10/11 pt Times by Setrite Typesetters, Hong Kong
Printed by Koon Wah Printing Pte Ltd, Singapore

Contents

Maps   vi
Tables   vii
Figures   viii
Illustrations   ix
Acknowledgements   xi
Preface   xii
Introduction   xiv

Part I THE MINING COMMUNITY
1 Development and change in resource-based communities   3
2 Local government and the Australian mining town   13

Part II NEW TOWN IN THE URANIUM PROVINCE
3 Uranium discoveries and development choices   33
4 Jabiru   54

Part III LIVING IN JABIRU
5 A community profile   97
6 Problems and prospects: the community response   135
7 Conclusions   158

Appendix A Writings on the Alligator Rivers Region   171
Appendix B Survey methodology   173
Bibliography   179
Index   193

Maps

1 Location of Jabiru   xvi
2 Selected resource-based towns in Australia   14
3 The Alligator Rivers Region, Northern Territory   35
4 The Ranger anomalies   39
5 The Ranger site   40
6 Possible locations for the town of Jabiru   44
7 The site for a new town   45
8 Jabiru Structure Plan, 1973   48
9 Kakadu National Park   52
10 Jabiru Town Plan, 1978   57
11 Jabiru Amended Town Plan, 1979   62
12 Jabiru Final Town Plan, 1979   63
13 Distribution of housing types in Jabiru, 1984   67
14 Distribution of single persons quarters, Jabiru   122

Tables

4.1 Cost estimates for different housing distributions   65
4.2 Ratepayers in Jabiru 1984   85
5.1 The people in Jabiru   102
5.2 Heads of households   103–105
5.3 The move to Jabiru   109
5.4 Satisfaction with housing   116
5.5 Reaction to dispersal of single persons quarters   123
5.6 Rating community facilities and services   126
5.7 Satisfaction with the community   131
6.1 Community problems and issues   137
6.2 Kakadu National Park   144
6.3 Aborigines in Jabiru   147
6.4 Community involvement   152
6.5 Assessment of governance bodies in Jabiru   154
B.1 Sample strata and response rates   176
B.2 Approximate sampling errors   178

Figures

1.1 Stages of resource town development   5
2.1 'Normalisation' concepts of Mt Newman Mining Co.,
    May 1983   25
3.1 Ranger uranium discoveries chronology   37
3.2 Organisations represented at the Ranger Uranium
    Environmental Inquiry   49
4.1 Jabiru planning and construction chronology   56
4.2 Chronology of events leading to local government in
    Jabiru   79
4.3 Chief bodies concerned with development of Jabiru,
    1983   88
5.1 Union protest notice displayed during the strike in Jabiru,
    November 1983   100

Illustrations

Aerial view of Jabiru, July 1984   xxvi

Housing types   68–69
High-set house, Jabiru East
Ranger house, Jabiru
Undeveloped subdivision (vacant lots), Jabiru

Aborigines in Jabiru   72–73
Restricted entry sign to Manaburduma Camp, Jabiru
Aboriginal shelter, Manaburduma Camp, Jabiru
Sign, Kakadu National Park

Jabiru town centre under construction   75

Jabiru town centre   76–77
Shops
Tourists arriving at Jabiru town centre
Kakadu tourist congestion at Jabiru town centre

Picket line on the road to Jabiru East, November 1983   98

Housing types   114–115
Ranger townhouses, Jabiru
Jabiru East high-set house rebuilt on ground in brick veneer in Jabiru
Government family house, Jabiru

Single Persons Quarters (SPQs)   120–121
Ranger SPQ with six units per building
Government SPQ with three units in this building
Front view of government SPQ

Jabiru town facilities   128–129
Child care centre in town centre
Olympic-size swimming pool with picnic area
Jabiru Cafe in town centre

Kakadu National Park   142–143
Town sign outside Jabiru
Jim Jim Falls in the Dry
Four-wheel drive track near Jim Jim Falls

Acknowledgements

The publisher and authors would like to thank the following for permission to reproduce these materials:
Aerial photograph p xxvi by courtesy of the Survey and Mapping Division, Dept of Lands, Northern Territory; (Jabiru town centre under construction photos) Jabiru Town Development Authority; (Map 4) Institution of Mining and Metallurgy, London, for Figure 1, p 297, in G. R. Ryan 'Ranger 1: a case history' *Uranium Prospecting Handbook* 1972; (Map 5) Australian Government Publishing Service, for Map 12, p 83 in *Ranger Uranium Environmental Inquiry Second Report* 1977; (Map 6) Australian Government Publishing Service, for Map 4-2a, p 36 in *Alligator Rivers: A Regional Study to Determine the Feasibility of Establishing a New Town in the Alligator Rivers Region of the Northern Territory* 1972; (Map 7) Australian Government Publishing Service, for Map 4-4, p 38 in *Alligator Rivers: A Regional Study to Determine the Feasibility of Establishing a New Town in the Alligator Rivers Region of the Northern Territory* 1972; (Map 8) Australian Government Publishing Service, map on p 38 in *Alligator Rivers N.T. Regional Town Project, Design Study—Stage 1* 1973; (Map 10) Centre for Resource and Environmental Studies, Australian National University, map on p 95 in *Social and Environmental Choice: the Impact of Uranium Mining in the Northern Territory* 1980; (Map 11) Centre for Resource and Environmental Studies, Australian National University, map on p 97 in *Social and Environmental Choice: the Impact of Uranium Mining in the Northern Territory* 1980; (Map 12) Jabiru Town Development Authority; (Figure 1.1) Government of Canada, table on pp 12–13 in J. A. Riffel *Quality of Life in Resource Towns* 1975.

Preface

This book has its origins in a research project funded by the Australian Research Grants Scheme. The initial focus was restricted to an examination of governance in the new town of Jabiru, NT, but later widened, with the support of the Jabiru Town Development Authority and the NT University Planning Authority, to enable us to undertake a book-length study. Researchers privileged to work in a small community have a responsibility to respect the privacy of their informants and we have tried to minimise attributable references to individuals where this might cause offence. Ultimately, the responsibility for what we have said and any errors or omissions are ours alone. While the workload has been shared evenly over the course of our research, it is appropriate to note that John Lea was primarily responsible for writing the Introduction, chapters 1 to 4 and the Conclusion, and Bob Zehner for chapters 5 and 6 and Appendix B.

We are grateful to many persons and organisations too numerous to list individually. However we make special mention of Geoff Stolz, Chairman of the Jabiru Town Development Authority whose active support was invaluable; Peter Loveday and staff of the ANU's North Australia Research Unit in Darwin who generously assisted us in overcoming the many practical obstacles associated with undertaking field research in the 'Top End'; Don Woods, Alan McIntosh, Bernard Fisk, Geoff Allen, Phil Baily and Neville Sheringham from mining companies in the Uranium Province; Eric Main, Chairman of the Jabiru Town Council and Councillors David Green, Kaye Danielson, Norm Tenthy and Gil Court, as well as members of the former Advisory Council; Kevin Witt, formerly Town Manager of Jabiru; Dan Gillespie and staff of ANPWS; Elspeth Young, Owen Stanley, Will Sanders and Rob Freestone who all gave us much needed support during the writing process; John Owen for cartography; and our survey assistants Debra Hinton, Wayne Houston and Marie Brink. Research of this kind also depends on the goodwill of numerous families prepared to put up with the inquisition of household questionnaires,

and to the population of Jabiru who fell within our sample we owe
a special debt of gratitude. Our wives, Dee and Ruth, made our frequent absences in the Northern Territory possible over a period of
three years and accepted a great deal more work as a result.
We also wish to acknowledge the financial assistance of the Northern Territory University Planning Authority which provided funds to
cover some of the publication costs of this book.

Introduction

> As communities beyond the city become increasingly important with growing national emphasis on resource development, the question of their governance (and public participation) becomes more salient. How much autonomy will they have, how much should they have if they are to contribute to the creation of the good society? (Bowman, 1981: xxvii).

The immediate reason for writing this book is to examine some of the events, decisions and paradoxes which accompanied the recent development of the Uranium Province of the Northern Territory (NT), one of the richest and most controversial mineral prospects in the world. An underlying impetus also arose from the realisation that little published work deals with the institutional mechanisms whereby new resource-based settlements are established and the way many of them are subsequently transformed into self-governing communities.

The special properties and uses of uranium and the location of our case study in a region recently included in the World Heritage List, have guaranteed a continuing and controversial milieu unmatched in the history of Australian mineral development. In spite of this dramatic context, however, the book is not about uranium mining nor the archaelogical and environmental wonders of the Kakadu National Park. It is concerned with the newly arrived population of miners and public servants who have been attracted to settle in the region. It examines the genesis of the new mining town of Jabiru and the emergence of a community under conditions which are a caricature of many of the contradictions confronting resource development in Australia today.

Background and objectives

On the national stage the search for new sources of energy has come up against a climate of opinion seriously questioning any use of nuclear fuels and increasingly aware of the fragility of the natural environment. In northern Australia and in the Uranium Province of the

Northern Territory in particular, there are the added dimensions of Aboriginal ownership of potentially rich mining land and a strong regional desire to quicken the pace of development. The uranium discoveries of the Alligator Rivers Region (ARR) offer an enormous boost to the economic potential of this remote region as well as to the economy of the Northern Territory itself. The effects of these underlying cross currents and interests have coloured the events surrounding the birth of the new town of Jabiru and make it difficult to view the experience of its early years dispassionately.

The range and severity of social, economic and environmental conflicts appear extreme but they are not unique and many of them can be expected to re-emerge as other large mineral discoveries are exploited in the years ahead. The town of Jabiru is located some 230 km east of Darwin (Map 1) on the borders of Arnhem Land and has become the population centre for the ARR. The latter corresponds approximately to the area covered by the catchments of the East, South and West Alligator Rivers and should be viewed in relation to two other important areal sub-divisions: the Kakadu National Park, which extends beyond the ARR boundary to the west; and the Uranium Province itself, which is a convenient term for the whole of this part of the Northern Territory where traces of the mineral have been discovered.

The background and origin of the new uranium mining industry is well documented due to the lengthy environmental impact investigations of the mid-1970s known as the Ranger Uranium Environmental Inquiry (Fox et al., 1976; 1977). The Inquiry's recommendations were generally followed by the commonwealth government and had the effect of limiting initial mining activity to one company and the status of the town to a 'closed' community serving the direct needs of the uranium industry alone. The first residents arrived in 1980 when production commenced at the Ranger mine and the town now has a permanent population of some 1200 people. The original plans were for a town of more than double this size but the failure of two prospective mining groups, Pancontinental and Denison Australia, to gain permission to export uranium from their leases at Jabiluka and Koongarra has resulted in a 'Ranger only' town. In spite of this, Jabiru still ranks as the sixth largest town in the Northern Territory and has assumed de facto status as the regional centre for the ARR. We stress the description 'de facto' because the town's 'closed' status has not been officially lifted even though it has been progressively eroded over the past four years. The fact that Jabiru was planned in this way helps to explain its appearance as a predominantly 'white' enclave in a region with a permanent Aboriginal population.

The region as a whole, 'with its uneasy blend of uranium, conservation, tourism, [and] Aboriginal rights' (Von Sturmer, 1982: 69), has been subjected to much detailed investigation over the past decade but

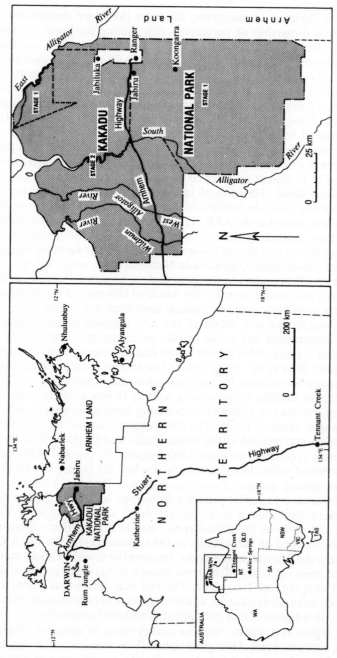

Map 1 *Location of Jabiru.*

there has been little comment on the town itself and the ways in which the residents are adapting to their novel surroundings. In fact the divided intergovernmental responsibilities for the town revealed in the wake of the Ranger Inquiry are a bureaucratic labyrinth (Lea and Zehner, 1985). Its development was to be controlled by two different statutory bodies: ultimate responsibility lay in the hands of the Director of the Australian National Parks and Wildlife Service (ANPWS) in Canberra who had, under the provisions of the *National Parks and Wildlife Conservation Act* of 1975, granted a 40-year lease over the townsite to the Jabiru Town Development Authority (JTDA). The latter had been set up under Northern Territory legislation in 1979 to construct and manage the town. Thus the new settlement was to be built under the control of the National Parks Act on commonwealth land, constructed by a Northern Territory statutory authority—and largely paid for by three prospective uranium companies! Moreover, most of the surrounding land had been transferred from Crown to Aboriginal ownership under the provisions of the Northern Territory *Aboriginal Land Rights Act* of 1976 and leased back to the commonwealth as a national park. It is hard to conceive of a less promising foundation for the eventual introduction of local government.

Thus our primary interest and the subject which forms the theme of this book is the origin and roles of these extraordinary institutional arrangements which, despite their complexity, were deliberately fashioned to involve all major parties to the uranium developments and, we will show, offered the future possibility of some measure of self-government. We were particularly interested in the progressive transfer of administrative functions from the Development Authority to elected town council from 1982 to 1984 and the reactions of residents to the planned introduction of some local democracy. Before outlining the Jabiru research project, however, it is important to identify the four main actors or bodies who were instrumental in setting up the bureaucratic controls over the development of Jabiru and who have each played significant roles to date.

The commonwealth occupied the central position, given its historical administrative responsibilities for the Territory (until self-government in 1978) and its control over the export of uranium products. The commonwealth's presence today is seen in the activities of its major agencies, notably the National Parks Service, the Office of the Supervising Scientist, and the Co-ordinating Committee for the Alligator Rivers Region.

Self-government in 1978 saw the new Northern Territory government inherit responsibility for creating the town of Jabiru. The *Jabiru Town Development Act* of 1979 was a crucial step in determining the character of the institutions which were to emerge in the town. The

Northern Territory administration was represented directly by the appointment of senior Darwin public servants to the Development Authority and through the operation of its various agencies and instrumentalities in the construction of the town.

Energy Resources of Australia Ltd (ERA), the main uranium mining company and owners of the Ranger mine, was formally represented on the Development Authority together with two other prospective companies. Although Pancontinental and Noranda (replaced by Denison Australia in 1980) possessed only guest status on the board, they were to influence many decisions taken during the initial construction period. Together with ERA, their position as paymaster (or eventual paymaster when the cost-sharing agreement between the companies was complete) gave them the dominating position in the Development Authority.

Aboriginal organisations represented by the Bureau of the Northern Land Council (NLC) and the Gagudju Association of local landowners were the fourth and weakest of the primary actors but they had considerable potential influence through the special status accorded them in the Ranger Inquiry recommendations. The NLC was actually offered two seats on the Development Authority Board in 1979 (*NT News* 21 November 1979) but declined to accept them in a decision they were later to regret (Lea and Zehner, 1985).

We turn now to the origins of the Jabiru project and outline the practical issues and difficulties we faced when embarking on this research.

Our investigations in Jabiru

Our initial contact with the town of Jabiru came in February 1982 during a fact finding visit to remote Australian mining towns. We had received funding the year before from the Australian Mineral Industries Research Association (AMIRA) to assist us in examining the social and physical planning problems of mining towns from the standpoint of the companies and government bodies responsible for managing them, the first AMIRA grant to be awarded for social research (Zehner and Lea, 1983). We soon established that there was a broad range of governance types in these towns about which standard works on local government have very little to say.

Early in 1982, the possibility of achieving community participation in town management appeared to be at the centre of an important debate in Jabiru and our decision to seek research funding to monitor its introduction was based on practical and academic grounds. The word 'appeared' is used advisedly because our initial contact was only with senior figures in company, government and town circles and did not include a cross-section of residents. We were later to discover that only a small proportion of the residents were likely to have been involved in the debate and that no more than a quarter of eligible residents voted in the council elections of 1982 and 1984.

We subsequently received financial backing from the Australian Research Grants Scheme late in 1982 and this allowed us to begin detailed fieldwork. It was important to secure this independent source of support; to have looked solely to the primary interest groups involved in Jabiru could have prejudiced our research design and conclusions. We later gained some assistance from our universities and from sources in the Territory (the Jabiru Town Development Authority and the NT University Planning Authority) to conduct a community survey, but it is unlikely this would have been forthcoming if we had not been able to establish a prior and independent presence in the town.

The modest level of funds initially available to us for the study ($11 600 spread over two years) precluded any extensive participant observation, although we did make a dozen visits to the town between 1982 and 1985, with a month-long stay during the survey in July 1984. This method of data collection was suited to the monitoring process because of the ongoing uncertainty attached to the timetable for the introduction of local government in Jabiru. As it turned out, the changeover date of 1 July 1984 had been sought three years before by Paul Everingham, then Chief Minister, and events were to exactly match these instructions. However, a number of changes among the key individuals concerned took place within this period and we tape-recorded interviews with this in mind.

Thus far we have outlined the background to the study but have said nothing about our own research interests. The individual imprint which researchers bring to community studies is often overlooked, as is the necessity to provide some comparative basis for the work. Bell and Encel's *Inside the Whale* (1978) and Wild's *Australian Community Studies and Beyond* (1981) have explored some of these issues but, more recently, Austin (1984) and Peace (1985) have called for Australian-based social researchers to examine more thoroughly the international content of their work with particular emphasis on historical specificity and the broader political and cultural forces present in Australian experience and society. Austin has also attempted to relate the main research traditions in Australian sociology in her pursuit of a more integrated approach. The four areas identified by her— Marxist class analysis, the quantitative examination of social mobility and status, community studies, and the role of the state—are relevant to our investigations in Jabiru and we have borrowed insights from all of them.

We both have the experience of teaching in town planning schools but our disciplinary backgrounds and previous fields of research are quite dissimilar. Lea was trained as a geographer and town planner in Britain and his doctoral research was in the field of African studies. His particular interests prior to the Jabiru project were in Third World and Aboriginal housing policy (Lea, 1979; 1982; 1983). Zehner is a sociologist with North American training in survey techniques and analysis (Zehner, 1971; 1977; Zehner and Chapin, 1974), much of it with specific reference to attitude studies in New Towns. The combination of these different backgrounds has resulted in an approach which aims to relate the Jabiru experience to a larger interdisciplinary literature than is usually found in Australian community studies. At the same time, the intention is to provide enough information to enable an empirical analysis of issues which have not been examined in any depth in Australian mining towns before, as well as making possible comparisons with case study material from elsewhere.

In our view, a full understanding of the present social and economic circumstances of Australian mining communities demands contributions from several areas of specialisation: the perspectives of political economists at the macro-scale, urban history, as well as local level analyses of social and institutional change. We concur with Austin's (1984: 185) assertion that:

> ...the reference point for analysis should be an understanding of Australia's colonial past and an appreciation of the likely course of its own short-term future. The prolific writings of political economists and economic historians in Australia offer the sociologist plenty of material with which to construct this reference point which should give equal attention both to class and institutional analysis. This will not be a glamorous sociology seeking to cast the intellectual in the role of major political actor. It will be one devoted to discovering Australia's past in its present, one trying to determine the major conflicts and changes likely to occur in the near future. It will be one which seeks to understand better the structure and culture of urban life, and the social relations of Australians in the industries of the rural north and west.

The framework we have chosen for our analysis of developments in the Uranium Province is discussed in detail in the following chapter. An immediate issue which demands our attention, however, is the complex nature of the uranium question and how it underlies and influences so many development decisions in the Alligator Rivers Region. Certain paradoxes, in particular, were contained in the commonwealth government's decision to allow mining to proceed (Commonwealth of Australia, 1977).

The uranium question

The whole development of Jabiru is overlain by constraints and uncertainties concerning the uranium question. Besides the higher than usual level of doubt about the future of the town which they have caused, there is the paradox of a large and enormously profitable enterprise which is the major extra-governmental source of financial support for Northern Territory Aboriginal groups as well as the target of the anti-nuclear movement (Altman, 1983; Cousins and Nieuwenhuysen, 1984). Some idea of the economic importance of the Ranger mine can be gauged from the fact that one year's production in energy terms is equivalent to one-and-a-half times Australia's annual electricity consumption, or about 11.6 per cent of the Western world's output of uranium oxide in 1984 (*Sydney Morning Herald*, 27 September 1985). This yielded in the financial year to June 1983, for example, $62.8 m in tax; $12.5 m to Aboriginal interests; $9.5 m in wages and salaries and $33.5 m in shareholders' dividends. The town of Jabiru has also become a tourist attraction it its own right, rivalling the most famous attractions of the Kakadu National Park. Company sources indicate that some 20 000 people per annum, or about one quarter the tourists now visiting Kakadu, inspect the Ranger mine and town.

Such paradoxes clearly suggest that the closer people and governments become involved with the Uranium Province and its human problems, then the greater is the likelihood that moral questions surrounding the mining and use of uranium will be subsumed by local issues. This helps to explain why uranium mining is not under such critical attack in the Territory—or for that matter in other resource-dependent and conservative parts of Australia—as compared to the metropolitan centres of the south. It does not necessarily follow, however, that northern Australians, both white and black, are morally less responsible than those critics of uranium mining who are not involved with the industry in any personal or economic sense (not all residents of Jabiru support uranium mining as we demonstrate in the concluding chapter).

Another interesting side effect of the uranium debate is the manner in which the strategic importance of the mineral has affected national thinking on the impact of major resource projects in general and development policy in the region in particular. Saddler (1980a: 194) and others suggest that only the complex *political* pressures associated with uranium caused the Ranger Inquiry to be held at all: 'It is salutory to reflect that, had it not been for the generic issues associated with the uranium mining—nuclear weapons proliferation, radioactive waste disposal, nuclear reactor safety, and so on—the Inquiry may well not have been held'. Similarly, although the intention to create a National Park in the region had been announced as early as 1965, it was the threat of uranium mining which served to concentrate government thinking.

The *realpolitik* of resources exploitation in Australia was such that 'Prior to 1975, there was no public participation in decisions about the development of the Alligator Rivers...' (Saddler, 1980a: 194). This situation irreversibly changed with the advent of new commonwealth, state and Territory environmental impact legislation (Fowler, 1982)— against the background of the Franklin Dam controversy in south-western Tasmania and the huge copper, gold and uranium prospects at Roxby Downs in South Australia.

Before outlining the scope and organisation of this book, it is important to offer some insights into the extensive literature which has been written about the history and development of the Alligator Rivers.

The Alligator Rivers Region

It is not intended here to provide an exhaustive review and bibliography of research in the ARR which forms a virtual sub-branch of northern Australian studies. About 50 selected references on the prehistory, history and modern development of the region are collected together in Appendix A and sub-divided into a number of interest areas of relevance to this study. It is a sample of diverse literature bringing to mind the plea made by the social anthropologist Clifford Geertz (1963: xvii), in a similar situation, that he be allowed to 'pursue scientific quarry across any fenced-off academic fields into which [he] may happen to wander'.

There was a human presence in the Alligator Rivers Region long before the arrival of the first white settlers in the nineteenth century. One of the fascinations for the observer of contemporary events is the fact that Aborigines have lived in this area for more than 20 000 years. The evidence of early settlement uncovered by anthropologists is a reminder that the Kakadu National Park is an inseparable amalgam of human and natural elements. This is an historical symbiosis suggesting to Jones (1984: 107) 'that the landscape to a certain extent has been an artefact of all those millenia of Aboriginal occupation and use'. Even in the modern era the link with ancient times remains. It was captured in somewhat mystical fashion in a newpaper article written at the height of the uranium exploration boom:

> They say that the mountain is inhabited by Dadbe, the King Brown Snake. The local tribe that inhabited the area is now all but extinct; but the name and the legend are feared by all the tribes of Arnhem Land. They believe that if the rocks of the mountain are disturbed the giant snake will emerge and destroy everyone in the world (*The Australian Financial Review* 16 July 1973).

The mountain refers to Mount Brockman, which contains the sacred sites of Dadbe and Djidbidjidbi and at whose base were found some of the richest uranium deposits in the world, the Ranger anomalies. The analogy between Aboriginal legend and military uses of the mineral was of course not lost on the journalist who wrote the piece cited above, and serves to remind us that in parts of the region the ancient symbiosis has given way to conflict. It is the human and institutional responses to these tensions and opportunities which really provide the overall setting for this book.

The Ranger Uranium Inquiry established by the Whitlam Labor government is the central focus for the policy-oriented literature. A number of conflicts over land use had been emerging well before the uranium discoveries were made (Saddler and Kelly, 1983: 268), such as the increasing antipathy between pastoral interests and environmentalists who had secured some official support for the creation of a national park. The Ranger Inquiry considered these and other issues in its wide terms of reference and was one of two important inquiries held under the provisions of the commonwealth *Environmental Protection (Impact of Proposals) Act* of 1974 (the other concerned sand mining on Fraser Island in Queensland).

In 1978, the commonwealth Minister for Aboriginal Affairs commissioned the Australian Institute of Aboriginal Studies (AIAS) to study such impacts and the results have recently been consolidated into a single report (Australian Institute of Aboriginal Studies, 1984). The social impacts of uranium mining on the Aboriginal community have produced a distinct group of publications. The work of individual project members is also available from other sources (Tatz, 1982; Tatz et al., 1982; Von Sturmer, 1982) and may be read in conjunction with some polemical critiques of the Aboriginal situation (Roberts, 1981; Howitt and Douglas, 1983), together with empirical research into specific issues such as the Gagudju Association (Stanley, 1982) and the question of royalties (Altman, 1983).

The Kakadu National Park has been the subject of research on visitor usage (Forbes and Merrill, 1983; Gale and Jacobs, 1984), tourism issues (Chaloupka, 1984; Jones, 1984; Palmer, 1985), and the involvement of Aborigines in park management (Weaver, 1984). The Park's management plan (Australian National Parks and Wildlife Service, 1980) is a good compendium of factual and policy material and is to be released in updated form covering the newly expanded park area.

Considerable information about the town of Jabiru is contained in unpublished submissions to the Ranger hearings and various planning reports for the construction of the town. But relatively little has been written other than earlier papers arising out of our own research (Lea, 1984; Zehner and Lea, 1984; Lea and Zehner, 1985).

Scope and organisation of the book

There are three parts to the book, each comprising two chapters. Part I sets the theoretical and historical context within which we have conducted our investigations in the Uranium Province. Chapter 1 looks at general aspects of the process of development and change in small single industry and mining communities. No single approach or discipline satisfactorily explains the social and institutional circumstances of these distinctive communities. Our analytical framework is also outlined here together with some hypotheses about the institutional arrangements which were created in the Uranium Province and the reactions of Jabiru residents to them and to the experience of living there. Chapter 2 looks at the way in which local government has been introduced into Australian mining communities in the past, comparing the early circumstances in the Victorian goldfields with the recent experience of the Pilbara iron ore towns in Western Australia.

Part II covers what we have termed a first level of analysis and examines the events leading to the construction of Jabiru and the subsequent introduction of local government. Chapter 3 considers the interactions and conflicts among the chief interest groups responsible for the creation of the town and Development Authority and chapter 4 examines the construction period and internal changes in the town during the transfer of administrative functions to local government in the formative first three years in Jabiru.

Part III is based on our community survey of the residents at the time of the governance changeover in July 1984 and looks at their response to many of the issues raised in the earlier discussion. In addition, we examine their reactions to living in a planned community which contains several innovative design features, as well as to the special circumstances of the town's location in the Kakadu National Park and its dependence on the uranium mining industry.

The following chapter, on the theoretical background to our work, is directed chiefly to those interested in the wider sociological study of mining communities. The reader whose interests lie primarily in Australian mining towns may wish to start at chapter 2.

Aerial view of Jabiru, July 1984

Part I THE MINING COMMUNITY

1 Development and change in resource-based communities

Many aspects of community living are determined in whole or in part by decisions made outside the community by policies and procedures of state or national organisations, by state or federal law, and by developments in the national economy. It is extremely difficult to separate these aspects from more locally determined institutional functions without cutting some of the live connective tissue which most sociologists agree is essential to community structure and process (Warren, 1956:338).

Modern approaches to the study of communities are less inclined than in the past to sever the 'live connective tissue' of community life linking internally and externally generated sources of change. The aim should be, following Austin (1984:139), to achieve 'A method of regional studies which moves between ethnography and structural analysis avoids the sins of triviality, overly abstract theory, and the presumption that ordinary people do not know what they are doing'. There is less of a primary concern today with the effects of locality, giving rise to something of a problem in the study of remote resource-based settlements where an approach is needed which recognises, without reifying, the important effects of geography.

Unravelling the events leading to the construction of Jabiru and the subsequent changes in administrative arrangements which followed is best tackled by treating geographical factors as constraints rather than as determinants of social behaviour. Thus, 'people who live in mining communities do not act as they do *because* of the shape, size and location of the community, but they are influenced by it. Geography restricts choice and freedom of action' (Bulmer, 1978:298–9).

We have found it necessary to examine recent events in Jabiru and the Uranium Province at two broad levels of analysis, corresponding to interactions between the main actors involved and the reactions of residents to living in a peculiarly sensitive and uncertain environment. This bi-level approach helps distinguish the major objectives of the investigation, though it does introduce an artificial division between matters which generally preceded town construction and were largely

3 ·

external to it, and those which have been subsequently generated from within. The analytic framework chosen is discussed in some detail here but, before doing so, it is important to outline the extent of available literature on social change in mining communities and previous contributions to furthering understanding of the situation found in Australia.

Sociological approaches towards mining town development and change are categorised by Bulmer (1975) as ranging from Marxist perspectives emphasising the social relations of production (Dennis et al., 1969) to contributions from industrial relations theory (Kerr and Siegel, 1954; Salaman, 1971). In addition, several attempts at depicting stages of settlement growth (Lucas, 1971; Riffel, 1975) suggest a commonality of experience as mining towns and single industry communities pass through readily identifiable stages from construction through a period of transition to eventual maturity (Figure 1.1). Although attractively simple, there is no explanation *why* one stage is supplanted by the next nor when change occurs in anything other than a linear sequence (Bradbury and St Martin, 1982). Thus a complete picture of social change demands a combination of explanation and description and no single approach is able to cover the field adequately.

The important question of institutional and governance considerations peculiar to remote resource-based communities has tended to be all but ignored in the literature, although there are notable exceptions in the work of Atkins (1976) in Australia and Paget and Walisser (1984) in Canada. The particular circumstances of the governance transition when control over these communities passes from the mining company or development authority to local residents is examined by Thompson (1981) for the Pilbara towns and by Lea and Zehner (1985) for the Northern Territory Uranium Province. Given the obvious importance of the minerals industry to the national economy, Australian researchers have paid remarkably little attention to the government and social welfare of the small communities concerned. Exceptions to this generalisation exist in the notable work of the CSIROs Resource Communities Environment Unit, based in Melbourne, and in a small number of book-length studies.

Australian community studies and the mining town

The few titles in the list of Australian community studies of mining towns exhibit a wide variety of objectives and methodologies. Five books can be differentiated according to method of analysis employed and broad type of mine settlement considered. The earliest of the

Figure 1.1 Stages and characteristics of resource town development

	Economic characteristics	Demographic characteristics	Social characteristics
Natural or prediscovery	No economic activity or only hunting and fishing by native peoples	No population or only small bands of native peoples	Unpopulated or small, isolated native communities in limited contact with white society
Prospecting to survey	Short-term activity. Money spent 'outside'. Traditional native economy persists with some trade with whites	Short-term summer residents. Young men, no women. If there originally native people in industry	Isolated. Usually access by air only. Shack towns without amenities. Some contact with native peoples
Industrial and town construction	The first boom period. Mushrooming economic activity. Natives may be employed	Mostly single men. Some young workers with families. Very high turnover rates. Natives in minority; only stable group in population	Isolated, but easier access to outside. Trailer towns with basic amenities and 'pub', Signs of social problems among native peoples
Industrial operation and community improvement	Shift in construction from industrial to residential and commercial. More money spent in town. Falling off in employment of natives	Slowing rate of turnover. Increasing number of married workers. Native people a small minority	Improvement of housing and community facilities. Completion of roads and communications services. Reduced social problems among whites increased among natives
Industrial and community operation	Construction over, services established. Much of labour skilled. Few natives employed	Turnover rates reduced to 60%, Young married workers in majority	Amenities well developed. Few social problems among the whites, but boredom among wives. Natives on welfare. Marked stratification
Community diversification	Stabilisation of industry. Expansion of other services, especially government. Small manufacturing	Labour turnover stabilises at 35%. Young married in minority	Employment for wives available. Special programmes created, largely for native people
Community maturity	Diversified economic based. Limited opportunities for expansion	Balanced population structure in terms of age and sex. Low rates of turnover	Sense of community and 'belonging'. Whites and natives on welfare. Less racial tension

Source: J. A. Riffel, 1975: 12—13. Reproduced by permission of the Minister of Supply and Services, Canada.

modern examples, *Coaltown: A Social Survey of Cessnock* (Walker, 1945) contains the analysis of a wide-ranging questionnaire survey conducted during World War II in a town partly dependent on mining and located in a well populated region close to Sydney. While Wild's (1981:99) judgement that Coaltown is 'a rather sketchy and superficial account' is somewhat harsh, the book does not go far beyond the presentation of survey data.

Only one study deals with the modern circumstances of a true mining as opposed to mining-related town. *Open Cut* by Williams (1981) is based on her doctoral study of the coalmining town of Moranbah in central Queensland's Bowen Basin. This 'sociological study of work and family relationships among the working class' (1981:9) employs Marxist and feminist concepts and places great emphasis on the break with the tradition of 'naïve' empiricism in British and American sociology. The end product is highly conditioned by the determining influence of the initial concepts and there are few points of direct comparison with other Australian works of the community studies *genre*, such as *An Australian New Town* (Bryson and Thompson, 1972) or *Bradstow* (Wild, 1974). The latter utilise survey techniques and extensive participant observation to gain an understanding of community life rather than the class-related position of a single group.

Barrie's (1982) historical account of the short-lived Rum Jungle mine in the Northern Territory (Map 1) and the township of Batchelor to which it gave rise, is the only detailed treatment of a uranium industry-dependent community. It contains a valuable compendium of detailed information on the physical development of the town and anecdotal material covering the changing circumstances of a small community which has survived the disappearance of its primary economic base.

Two other works are concerned with remote communities in northern Australia and examine towns based primarily on mineral processing and port activities. They are *Dominance of Giants* by Stockbridge et al. (1976) and *The Hedland Study* by the CSIRO (Brealey and Newton, 1978). Both are located in the Pilbara of Western Australia, use extensive questionnaire surveys and do not employ participant observation in their research methodologies. The Hedland Study is an interesting variant on the sociological emphasis of the others in its consideration of town planning in the new suburb of South Hedland, designed to reflect 'Radburn' (pedestrian-vehicular separation) principles. This theme is continued in our study of Jabiru regarding certain aspects of the new town neighbourhood design.

Dominance of Giants possesses a great deal of tabular data, but also contains much perceptive comment about life in the Pilbara. Its purpose was to examine the quality of life in the region and the five small settlements of Roebourne, Dampier, Karratha, Wickham and Point Samson were chosen because of their good surface communications. The 'Giants' referred to are the four external caretakers of the Pilbara: the multinational mining groups, Western Australian and commonwealth governments and the trades union movement. It is the interaction among these powerful actors which were perceived to be the determining influence on almost all development decisions in the region, making the book an important forerunner to our own identification of similar influences in the Uranium Province.

This review of some major Australian studies of mining towns does not do justice to all the interdisciplinary work in the field but does reveal the paucity of community-based research in places where small populations belie their economic importance, geographical significance and unique characteristics.

The analytic framework

An analytic procedure was sought which could place the investigation in the context of the historical evolution of governance arrangements in Australian mining towns, the political and economic contradictions in resources development policy and the wider sociological generalisations about these communities. We were concerned to link the study of public policy and political priorities regarding Jabiru and the Uranium Province with an examination of community reactions to the choices made.

As noted earlier, two main levels of analysis were employed in our study. The first examined interrelations among the four chief actors involved in the uranium development decisions and plans to establish a national park in the early 1970s: the commonwealth government, Northern Territory administration (especially after 1978), the uranium mining companies, and Aboriginal groups. Most of our information was collected through interviews with key individuals in Canberra, Darwin and Jabiru during 1982/85 together with a search of documentary sources many of which were unpublished and unavailable in library collections. The second tier of investigation focused on reactions of town residents to the governance transition in Jabiru and related aspects of life in the town during the formative three-year period after construction was complete. Much of this was based on a questionnaire survey of more than half the households, undertaken in mid-1984, together with tape-recorded interviews with key personnel during the previous two years.

The central questions at the first level of analysis are straightforward: why was a development authority chosen to construct and

manage the town, and why was representative local government introduced so soon? Initially, this came in the form of an advisory council
in 1982 followed by a full town council two years later, in spite of
early opposition from the mining companies and lack of interest from
most residents. There are also underlying theoretical questions: do the
institutional arrangements chosen at Jabiru and the changes made
to them provide further evidence of the systematic co-optation of the
Australian working class (Connell, 1977) and changing priorities of
mining capital (Bradbury, 1979; Thompson, 1981)? Or, might we not
obtain a different and perhaps better understanding of influences in
the community through the study of some of the intergovernmental
relations involved and the competition and conflict inherent in the
development of this significant part of the Northern Territory. The
latter issues are obviously central to any contemporary analysis of
resources development but have tended to be subsumed in recent years
by more abstract and deterministic notions of the workings of
Australian capitalism.

In order to give some direction to the investigation we formulated a
working hypothesis which leans towards the second of the two prop-
ositions suggested above. Political commentators on the Northern
Territory scene have noted that the quickening pace of development
in recent years has begun a general move towards introducing local
government to the smaller centres wherever practicable: 'The rapid
growth of the population since the mid-1960s has brought with it dissatisfaction in town management boards formed in the early 1960s to
•advise the administrator. Local government is now being phased into
communities outside Darwin and services are provided before rates
are levied' (Chapman and Wood, 1984: 31, citing Heatley, 1978). The
new Everingham government maintained this policy on gaining office
in 1978 and swiftly introduced local government to the old settlements
of Katherine and Tennant Creek, as well as seeking its early presence
in the new towns of Nhulunbuy and Jabiru, and Palmerston just
outside Darwin. Although it was not to succeed in Nhulunbuy, we
may suppose that the directive from the Chief Minister to bring local
government to Jabiru by mid-1984 was part of the Territory-wide
trend.

On the face of it this supposition is not entirely satisfactory because
there are few points of comparison between the administrative demands of old diversified communities like Katherine and Tennant
Creek, the new Darwin satellite of Palmerston and the 'closed' mining
towns of Nhulunbuy and Jabiru. The latter were not experiencing
great population growth, were characterised by leasehold status of
town land and were not subject to much obvious dissatisfaction with
existing administrative arrangements. This demands that we advance a

better, or at least additional, explanation for the government's action.

An underlying reason why there was Darwin-inspired support for the introduction of local government to Jabiru was because of the Territory government's concern from the outset to secure a political presence and popular platform in a development region dominated by Canberra and company boardrooms in the south. Bauxite mining in the Gove Peninsula and uranium in the Alligator Rivers Region were effectively controlled by bodies external to the Territory and what little local influence there was seemed to be shifting progressively towards Aboriginal associations. The new and largely white residents of the mining towns in these rich development prospects, though small in number, were potential Territorians and formed a substantial community in Northern Territory terms. This was of great interest to power brokers in Darwin.

We have chosen to use a set of ideal types to assist us in the investigation of various intergovernmental tensions surrounding the competing interests in the Uranium Province. One of the clearest expositions of this approach comes from Saunders (1983:52) who states that 'The role of ideal types. . .is not to generate knowledge but to make knowledge possible by means of prior conceptualisation which enables us to classify phenomena and to sort out that which is relevant for our purposes from that which is not.' The four dimensions of the Saunders' (1984) model of conflict among different levels of government are:

1   An *organisational* tension of the kind surrounding 'central direction and sub-central autonomy and self-determination'. To what extent, for example, do appointed office holders or elected councillors in the Uranium Province have the ability to initiate and carry out their own policies in the face of controls from outside?
2   An *economic* dimension surrounding competing intergovernmental and interagency demands for expenditure, best illustrated by the unwillingness of the Northern Territory government to fund certain standard amenities in Jabiru because of its overall lack of jurisdiction in the Uranium Province;
3   A *political* dimension which Saunders (1984:89–90) defines as 'the relation between competitive, open or "pluralist" modes of interest articulation and more closed "corporatist" modes' or, put another way, the tensions between democratic accountability and corporate and efficient strategies. This is seen clearly in the differing objectives and managerial styles of the Jabiru Town Council and Aboriginal associations as opposed to corporatist agencies such as the National Parks Service and the Jabiru Town Development Authority;
4   An *ideological* dimension in the tensions between the rights of

private property and those of citizenship or between individualism and collectivism. This is best expressed in our context, perhaps, by special factors concerning the patterns of Aboriginal
and non-Aboriginal land ownership (and non-ownership) in the
Uranium Province.

We accept that some of these elements of governmental and agency
relations are less important than others in assisting us to come to an
understanding of the situation prevailing in the Uranium Province in
the 1970s and 1980s, and that they do not necessarily account for the
choices made nor the changes which resulted. The typology is not
intended to cover the activities of large private corporations such as
the mining groups and we have chosen to pay less attention to the important dealings of these companies with government than is covered
in other studies (Galligan, 1983).

The other primary objective of the study, categorised here as our
second level of analysis, was to monitor the progress of the governance transfer during the early years of the town together with reactions of residents to the changes. In addition, we hoped to assess
how residents responded to some of the peculiar effects of living in a
national park on Aboriginal-owned land. This situation has led to the
imposition of constraints over personal freedoms of a kind more far
reaching than normally experienced in 'closed' company towns—and
certainly foreign to the easygoing circumstances of life in the 'Top
End'. We adopted a proposition that it was likely that the push for
local government from within the town, such as it was, was from occupational groups outside the company workforce because of a wish
to gain a voice in community affairs dominated by mine management.
As far as the National Park was concerned, we thought it likely that
most residents, regardless of their employment, were in opposition to
the extra controls over personal behaviour imposed in Jabiru.

Although it does not cover the full extent of our investigations
in the town, we have adopted Bulmer's (1975) list of eight ideal
characteristics of mining communities as a means of illustrating the
social composition of Jabiru. The original typology is really aimed
at the traditional European rather than new Australian mining town
but, again, we consider the 'construction of abstract ideal-types rec-
ommends itself not as an end but as a *means*' (Weber, 1949:92).
Clearly then, *departures* from the ideal type can be important in coming to more accurate and useful generalisations. Bulmer's original
eightfold list is:

1   *Physical isolation* resulting in dispersed settlement and remote-
    ness with all the human implications of this situation. This is par-
    ticularly relevant to circumstances in northern Australia, though

Jabiru is not as remote as many other small mining towns;

2 *Economic predominance of mining* which relegates the residential settlement of the mine to the status of an adjunct to the single dominant economic activity;

3 *The nature of work*, suggesting the particular dangers and skills associated with mining and the precise working arrangements connected with this form of employment;

4 *Social consequences of occupational homogeneity and isolation.* The ideal notion is that the mining town consists 'exclusively of members of the manual working-class with the exception of mine management and shopkeepers and other service workers who live locally' (Bulmer, 1975:86). There may thus be little opportunity for upward social mobility within the community;

5 *Leisure activities* designed to contrast with the tight constraints of mine work. Occupational ties are seen reinforced outside of the work place;

6 *The family* exhibits special characteristics with the segregation of male and female roles, the former concentrating on activity outside the home and the latter within it;

7 *Economic and political conflict* demonstrated by the opposing interests of mine owner and worker in capitalist society. Employers will favour the status quo whereas workers will seek limitations on the ability of owners to exploit them. But, as Bulmer (1975:87) points out, '. . . conflict over the distribution of available economic resources will, in the pure capitalist mining community, entail the subjugation of the miner entirely to the interests of the mine owner, since the latter dominates not only the industrial sphere but all aspects of the economic life of the community'. Thus, opposition in the work place can be met by the denial of basic needs such as shelter and the right to bring visitors into the community who might help the cause of organised labour;

8 *The whole.* This notion includes the suggestion that there is something special about community life in mining towns which stems from the interrelationship of all the previous characteristics (for example, the strong mutual aid often noted in old mining communities, a feature witnessed during the long running coalminers' strike in Britain in 1984/85).

It bears repeating that the above set of abstract dimensions is an aid to empirical analysis of society in Jabiru and that it does not contain important elements of the situation such as the Aboriginal presence and the particular significance of uranium itself. It should be possible, however, to refine these dimensions to better describe conditions in the remote Australian mining town. The great variety of possible approaches towards the social analysis of these places requires a combination of investigative concepts and techniques tailored to the resources and time available. Our study of minerals-led urbanisation in the Uranium Province of the Northern Territory focuses on a settlement phenomenon with a history in Australia dating back to the mid-nineteenth century. We turn now to some of the recurring community issues which have characterised the creation of new mine settlements and, in keeping with our primary objectives, concentrate on the role of local government and the manner in which it came to be introduced.

2 Local government and the Australian mining town

> Castlemaine was sufficiently advanced to be erected into a municipality; a vast deal of money was invested in land and building here; and it was a duty the inhabitants owed to themselves to carry out local improvements and thereby stimulate their local trade, increase their social comfort, and enhance the value of their property (Melbourne *Argus*, March 1855, cited in Barrett, 1979:162).
>
> A historic multi-million dollar agreement in which a company will make a gift of an entire iron ore town to a local authority and residents is expected to be signed soon... The company wishes to see Newman as a 'normal' West Australian town, particularly with people beginning to buy their own land and homes (*West Australian Sunday Times*, 13 January 1980, cited in Thompson, 1981:317).

This chapter looks at two famous mining 'rushes' in Australian history which were both to lead to the creation of major new urban settlements and, eventually, to new municipal government in places located on opposite sides of the nation. They are events separated in time by 100 years, the one a tumultuous gold rush that was to attrack 100 000 miners from all over the world to central Victoria by 1855 (Blainey, 1963), and lead to the establishment of more than 20 new municipalities. The other, a carefully planned and privately financed operation, was to result in 50 000 new settlers moving to some dozen Pilbara iron ore towns and ports by the early 1970s. Here too, the workforce was international and, within a decade, local government was to emerge but in a rather different way.

Why should it be necessary or even useful to make a comparison of this kind, given the obvious contrasts between events separated by so much in time and differentiated in so many ways? The answer, as we suggested in the Introduction, lies in the need to examine Australian urban issues in their historical context and to emphasise the continuity of many problems facing communities at the local level. The town meetings at Castlemaine and Sandhurst (later renamed Bendigo), for example, which were called in early 1855 to discuss a petition for the introduction of local government are not unlike similar gatherings

13

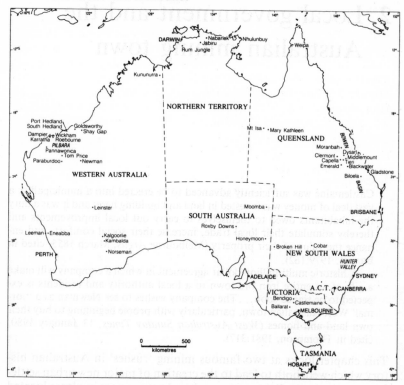

Map 2   *Selected resource-based towns in Australia.*

held in the company towns of Tom Price and Newman a century later (Map 2). For all the obvious differences, the arguments in essence were the same, a concern for private property, the stimulus to local industry and trade, the costly provision and maintenance of amenities, the fear of taxes, and so on. Above all, there seems to be a desire for some effective local autonomy and freedom from the controls of the mining companies, particularly among those employed by the non-mining sectors of the urban economy.

It would be wrong, of course, to overwork the comparisons noted here and to ignore the different governmental and economic circumstances affecting the genesis of local government under pre-federation and modern conditions. The close cooperation between capital and the state which has facilitated access to huge international borrowings in order to fund modern mining prospects had not yet arrived in the earliest examples we examine. Nor for that matter were there large mining groups which were interested in the advance provision of settlements and other aspects of social overhead infrastructure. The nature of our information sources also complicates direct comparisons, having the effect of concentrating discussion at the urban scale in the pre-federation examples and at company and higher levels of government in the modern situation. In spite of these imperfections, however, our discussion of contemporary events in the Northern Territory demands a good understanding of the circumstances which have led to the introduction of local government in earlier Australian mining communities.

Local government in nineteenth-century mining towns

The richly documented history of urban settlement in Victoria also includes much information on the early gold mining towns (Serle, 1963; 1971; Bate, 1978; Barrett, 1979). They were the means by which urbanisation was introduced to many inland parts of the state in the mid-nineteenth century and Barrett (1979), in particular, brings this period of intense settlement to life in his description of the genesis of local government in the gold towns. The discoveries had an enormous impact on inland Victoria which can be gauged from the fact that it produced one-third of the world's gold in the 1850s. In addition, immigration accounted for almost 90 per cent of the large inter-censal population increase between 1851 and 1861 (Urlich Cloher, 1979).

The mining industry at this time was labour intensive and the Victorian gold rush which had begun in 1850 was to create intolerable living conditions in some of the camps. According to Blainey (1963:52) 'Over ten thousand men lived in Ballarat's straggling canvas camps towards the close of 1854, and the atmosphere was explosive.' The famous miners' riot at the Eureka Stockade which took place in December of that year quickened the pace of Victoria's political democracy in Blainey's view and, by coincidence, occurred in the same month in which the *Municipal Institutions Establishment Act* became law in the Victorian Legislative Council.

It appears, however, that the initial impetus which persuaded some prominent sections of society in the larger gold towns (such as Castlemaine, Sandhurst and Ballarat) to seek municipal status in the following year arose out of something else. This was a combination of recent threats of intervention by the state government to secure sanitary development in the towns, together with the personal objectives of some owners of private property and commercial capital. It could be argued that the former was a necessary condition before the well-being of the emerging gold mining industry could be secured and was therefore functional to the long-term interests of mining capital. The latter motives were an early illustration of the close connections between the fortunes of local government and the interests of the holders of private property. This is a matter to which we will return later in the chapter.

The new *Municipal Institutions Establishment Act* was modelled on the English urban district and provided a means for any interested community of no more than nine square miles containing at least 300 householders to seek municipal status on consideration of a petition containing at least 150 signatures (Serle, 1963:192–93). In addition, the *Public Health Act* of 1854 had given the state powers to introduce sanitary improvements and charge for them whether the town concerned possessed local government or not. This somewhat coercive legislative environment seems to have persuaded many leading citizens that their interests would be better served by their own taxing arrangements through local government than through interference of this kind from Melbourne.

It was not a unanimous view, however, and some twelve months of argument and counter-argument were to take place before the first elections in the gold towns occurred in January 1856. In summary, the arguments in favour of municipal incorporation seemed to have more to do with prospects for commerce and the economic life of the towns than pressing welfare and amenity considerations. There was particular interest, for example, in the size of a possible endowment from the state government (Sandhurst); in the stimulus to local trade which would accompany local government (Castlemaine); in the enhancement of property values (Castlemaine); and in the encouragement of more speculative capitalist enterprise (Sandhurst). The champions of the pro-municipal movement were those with primary commercial interests rather than involvement in the mining industry as Barrett (1979:170) observes in respect of the backgrounds of Castlemaine's first councillors, 'The seven elected were all storekeepers, all "highly respected and efficient men"'.

Those who opposed the introduction of local government seemed chiefly concerned about the impost of rates and taxes, wasteful public expenditure and 'jobbery', or the creation of jobs for friends and supporters. Although the membership of this group seems to have cut across occupational boundaries, there is some evidence that the owners of substantial grazing property and the mining interests were strongly represented (Barrett, 1979). In the case of Tarnagulla (a small mining town located between Maryborough and Bendigo) the opposing factions were to win and it provides a particularly clear example of the two contrary points of view:

> A 'progress' committee was formed at Tarnagulla in 1862 to seek a local municipality. The committee circulated a petition and called a public

meeting. The committee argued that Tarnagulla would become annexed to (and would be taxed by) another municipality; it was better that Tarnagulla should tax itself and spend the money locally. A committee member said that Tarnagulla had lost by inaction, 'having no recognised responsible body to advance their interests, while other districts owed their prosperity alone to the exertions made by their municipal bodies'. The committee was opposed by Tarnagulla's mining interests, who circulated a satirical post-card advising candidates for municipal office to apply in the year 1890. A spokesman for the mining interests said that 'doubtless a municipality would benefit the storekeepers and householders, but it would greatly in-jure the mining interests; the town would be in the hands of the council, and they would raise funds on the mining plant' (Barrett, 1979:243).

A variation of this argument was to be used by the mining company in Jabiru during the discussions about the possible introduction of local government there in 1982/83. In this case, as we shall see, the company argued that borrowings which it used for the construction of the town were secured against the headworks and mine plant, making a transfer of assets to local government undesirable.

In many of the larger gold towns, however, it was the supporters of local government who prevailed and more than 20 of the new munici-palities were proclaimed in the six years from 1857 to 1862 (Barrett, 1979). The volatile characteristics of their early years as mining camps extended in some cases into the initial period of local government, resulting in Castlemaine in 'ten elections, sixteen councillors, five town clerks, and three surveyors' (Barrett, 1979:173), all in the space of two years. The statistics for municipal construction in towns like Ballarat by the 1870s were equally impressive, with '56 churches, 477 hotels, 10,000 dwellings, 84 miles of streets, 60 miles of water mains, 50 miles of gas mains...' (Stacpoole, 1971:71). It is fitting, perhaps, to leave the last word to the novelist Anthony Trollope who visited the Victorian goldfields during this period: '...Gold brought to her the population which demanded and obtained that democratic form of government which is her pride' (Trollope, 1873:243).

As we have noted, the new councils received plenty of support from commercial interests but it still seems unlikely that they could have been so readily established in the face of direct opposition from the larger mining groups. There is some evidence in the case of Broken Hill in the neighbouring state of New South Wales that some big companies were indifferent to town affairs as far as their absentee management was concerned. Kennedy (1981:119) observes that BHP, the major company, located its directors in Melbourne who ran the mine from there in the late nineteenth century for the benefit of its metropolitan shareholders. 'Their *largesse* flowed into other pockets and they were in no position to lead local opinion on the life and death issues of water and self-government.' However, the composition of

Broken Hill's first council in 1888 indicates it was strongly supported by local mine management for, of the twelve elected aldermen, there were three mine managers, two mine employees, two mine agents, two building contractors, a publican, a boarding-house keeper, and a storekeeper (Kennedy, 1978:44).

The people of Broken Hill were united in their strong condemnation of what they saw as the town's neglect by the distant state government in Sydney and this focused opinion in a manner which was not as evident in the Victorian towns. Local government and even the threat of secession to South Australia became the means of expressing opposition to the very low levels of local autonomy which existed on the Barrier at this time: 'Only the crisis of water famine and threats of civil insurrection forced Sydney to act and set up municipal institutions in the town' (Kennedy, 1981:116).

Thus the origins of local government in Broken Hill were deeply affected by isolation, intergovernmental rivalry, intrastate tensions and absentee mine management, all factors which have persisted to some extent in the modern examples from northern Australia to which we now turn.

Local government and company towns in the Pilbara of Western Australia

A new climate for large-scale mining emerged in Australia in the 1960s with the attraction of huge amounts of multinational capital to fund the exploitation of mineral prospects in remote parts of the north and west. The postwar expansionary boom in advanced industrialised economies fuelled a demand for iron ore and energy which for supplier countries like Australia could not be met by drawing on local financial resources alone. Some of the urban effects of this influx of foreign capital and the changes it was subsequently to bring to metropolitan centres such as Perth (Harman, 1983) and Sydney (Daly, 1982) were to be seen in more immediate fashion in the 'instant' construction of company towns in the Pilbara iron ore region of Western Australia.

This modern equivalent of a nineteenth-century gold rush was made possible by a close partnership between domestic and foreign mining capital and the West Australian state government (Harman, 1982), which placed the responsibility for most of the construction and servicing of the new towns in private rather than public hands. Changes in government and company policy towards the Pilbara towns make an interesting modern case study of the coming of local government to new mining towns and, as we shall see, bear some comparisons with the earlier examples. The circumstances in the Pilbara towns are first of all considered in terms of the conditions and objectives which prevailed at the time of their establishment in the 1960s; second is an examination of the state government's review of the administrative arrangements in the towns which occurred a decade later; and third is an assessment of the radical critique of government and company behaviour which was directed at the transfer of some of the towns from company to shire control in the early 1980s. The latter critique became part of the 'normalisation' debate and is an example of the use of Marxian political economy notions of mining town development and change.

Establishment of Pilbara company towns

The state government of Western Australia became actively involved in the removal of existing constraints on resources development in the 1960s, such as the commonwealth embargo on iron ore exports. This alone would not ensure large-scale exploitation without the attraction of substantial amounts of foreign capital (Layman, 1982) and it was an innovation which did not receive wholehearted bi-partisan support in parliament:

> Mr Court: The honourable member referred to the fact that he was not convinced that behind the scenes...there might not be some Japanese capital. Let me make it clear, if there is, then what of it? Here is a nation which is our biggest customer today.
>
> Mr Tonkin: I prefer Australian capital.
>
> Mr Court: Would not we all? Tell me where we would get £30,000,000 in Australia, of Australian capital, to start from scratch a project of this nature! We would be laughed at. We have found the people who can provide this sort of money. They cannot take away the port; they cannot take away the railways; they cannot take away the towns; and they cannot take away the homes which will be developed (Government of Western Australia, 1963b:1680).

This was a persuasive argument at the time but, as Harman (1983) has recently suggested, the enthusiastic support for a major influx of mining capital may also have marked the entry of the state's economy into what has been termed elsewhere as the 'staples trap'. This signifies a distortion of the economy towards the extraction and export of raw materials which might prove difficult, or even impossible, to counteract. It is also a condition which progressively ties the interests of the region concerned to those of multinational companies.

The first of the iron ore agreements was signed between the state government and a consortium of companies comprising Goldsworthy Mining Ltd and three multinational partners in September 1962, and led rapidly to similar agreements with three other groups of joint venturers. Among the various conditions contained in the Goldsworthy agreement was a clause requiring the companies to 'lay out on the mining area a site for a town and in relation thereto provide adequate and suitable housing school roads amenities and water power and other services as shall be reasonable having regard to the locality' (Government of Western Australia, 1962a:56). The novelty of this form of mining town establishment did not escape the attention of the Opposition member for the Pilbara in the state parliament, who observed: 'I am not one who feels over-happy about public utilities as we know them being in the hands of private companies. I think complications do arise from time to time, and I think that Governments are responsible for those facilities' (Government of Western Australia, 1962b:648).

The precise nature of some of the possible 'complications' was not revealed until the state parliament considered the Hamersley Range Agreement a year later (Government of Western Australia, 1963a). In that debate the Member for Boulder-Eyre was much more specific about the actions which the companies might take in evicting troublesome or redundant tenants, who would then suffer the hardship of having to leave the area altogether with their families. He also raised some of the industrial implications which could flow from company ownership of the towns and drew on previous experience in such matters to support his argument (Government of Western Australia, 1963b:1671).

The reservations expressed were an accurate insight into some of the more obvious and extreme difficulties associated with 'closed' company towns and a pointer to some of the social contradictions which were later to emerge. There was little suggestion at this time that the government would recognise the future implications and, indeed, the response of Charles Court, the Minister concerned, could have come straight from the joint venturers themselves:

> It is a peculiar situation where the company has to provide all its own housing, and it has been customary in agreements of this type for certain rights and privileges to be given to the company so that it can *protect the workers themselves*. . . Unless the company has access to its own housing it can get into the position that somebody who has been dismissed for a good and sufficient reason—probably a reason which his own colleagues would think was fair and proper—hangs on to a house which should properly be made available to the new employee (emphasis added, Government of Western Australia, 1963b:1678).

It was a paternalistic attitude which was soon to be turned on its head by the companies in a policy reversal which some commentators were to suggest had little to do with the personal interests of employees. Before moving on to address this matter in the review of town administration which was to occur in the mid-1970s, it is important to note there had been an earlier new mining town in the Pilbara region which pre-dated the iron ore agreements by some 20 years.

The opening up of the rich blue asbestos discoveries in the Wittenoom Gorge of the Hamersley Ranges shortly after World War II by CSR Ltd, had been facilitated by the provision of a new town paid for by the state government (Layman, 1982). It was a development which has some interesting parallels in the analysis of the situation in Jabiru 30 years later. A senior West Australian public servant formally suggested a very similar model for the town of Wittenoom. It was a piece of advice which went unheeded at the time in Western Australia but contained an argument which was to be recognised after the uranium discoveries in the Northern Territory. The suggestion made in 1947 was that a small executive committee comprising the government departments most involved in the project be established as a 'controlling or organising authority to co-ordinate all activity' (Layman, 1982:155). Construction of the town should be quite separate from development of the mine itself.

Nineteen forty-seven was, coincidentally, the same year in which the New Towns legislation was enacted in Britain establishing development corporations as the primary means of implementing construction of these new communities. As we shall see, the development authority model which emerged 30 years later in the Northern Territory at Jabiru was to go considerably further than the suggestions for Wittenoom and included the mining companies as major participants. The important principle advanced in 1947, however, separating the development of the town from the mine, was not accepted in Western Australia until the review of administrative arrangements in the Pilbara towns of the mid-1970s.

The 'Carly Report' of 1977

By late 1976, the state government had begun to question the wisdom of continuing with 'closed' company towns in the Pilbara (Thompson, 1981), realising that this form of management effectively prevented diversification of the economic base as well as being a barrier to the entry of other industries. Accordingly, a former commissioner of Roebourne Shire, P. L. J. Carly, was appointed to investigate and recommend 'actions to progressively establish the normal roles of local authorities and the communities in nominated mining towns throughout the State' (Carly, 1977:2). The 'Carly Report', as it became known, was submitted to the Minister for Industrial Development in August 1977 but was not tabled in the parliament until two years later. As Thompson (1981) has shown, background information contained in the Report revealed some things about the original strategy to build the Pilbara company towns which were not generally known at the time. For example, it seems that bargains were struck between the companies and the state which agreed a lower rate of royalty payments in return for company development of the towns, 'In short it was the only way to cope with an emergency situation, but it should be remembered that it was not necessarily a net financial gain to the State since other considerations of the arrangement would automatically adjust accordingly' (Carly, 1977:3).

The original plan was for reports to be prepared and tabled for the towns of Wickham, Dampier, Newman, Tom Price, Paraburdoo and Pannawonica but only volume 1, containing general recommendations, and volume 2 on Wickham were made available to the public. According to Thompson (1981:306), a Murdoch University academic, his own action in distributing critical comments on the Report to unions in the Pilbara resulted in secrecy on the part of the state government and an embargo on the other volumes. A series of meetings and negotiations were subsequently held between the Department of Industrial Development, the shire councils of West and East Pilbara Shire and the Shire of Roebourne, as well as two of the mining groups with towns in these local government areas (Hamersley Iron and Mount Newman Mining). The first communities to be involved were Tom Price, Paraburdoo and Newman and the events leading to the incorporation of Newman in July 1981 provide a good example of the main elements in the normalisation debate.

The government had decided that the first stage in the negotiations should commence with agreement being reached between the local authority and mining company concerned (personal communication: John Read, former Shire Clerk, East Pilbara Shire, 1983). As Thompson (1981) points out in the fullest available discussion of the negotiations, this procedure was to prevent residents in the company towns from becoming involved effectively and gave control over the transfer to the companies and the state Department of Industrial Development. The latter bodies had much greater technical and financial resources than the small rural shires in the Pilbara and were able to present a *fait accompli* to them for their agreement. In the case of Hamersley Iron's negotiations with the West Pilbara Shire over the towns of Tom Price and Paraburdoo, Thompson (1981:309) observes 'The way the Agreement appeared to a number of councillors is that Hamersley Iron Pty Ltd was hiring the Shire to run the towns, take responsibility for complaints, *and deflect criticisms against Hamersley by workers on issues of town amenities*' (emphasis in original).

In Newman, the secretive nature of the negotiations relegated the involvement of residents to a town meeting in August 1980 entitled 'What does normalisation mean?', by which time most of the hard decisions about matters such as the sale of company housing and the future financial status of the community had already been decided (Thompson, 1981). The words of a union convenor recorded at the end of the Newman meeting demonstrate the feelings of some of the 100 residents who were present:

> I don't believe it! They say if we don't like it we can leave. But if nobody lived here who would the Government and the company insult or manipulate? We do live here and so we have the right to make decisions about our community and the quality of our lives without being insulted, manipulated or treated like children. It is *not normal* for the Government and company to make all of the decisions and then invite us in to ask questions when it is over (Thompson, 1981:315).

In order to present a balanced viewpoint, it should be acknowledged that the matter had been publicly aired in the state parliament some 14 months previously during the debate on the *Iron Ore (Mount Newman) Agreement Act Amendment Bill*. This measure was generally supported by both sides of the House and provided for changes to the original agreement which would allow the introduction of a new company employee house purchase scheme, an impossibility under the conditions of a closed town. The Opposition did indicate that they felt the changes were being introduced without sufficient consultation in the towns and raised several specific and potential difficulties with the housing scheme and the normalisation programme. Among them was the poor financial condition of the Pilbara shire councils and the fear that they would not be able to maintain the high physical standards in the towns or pay for necessary additional staff. This was an issue which was to re-emerge later in the negotiations, among accusations that the mining companies had obtained a very good deal from the state government.

Rural local authorities in Western Australia have been categorised as weak, 'emasculated shadows', in comparison to those in other resource-rich states such as Queensland—'the Local Government Act has about 300 sections that refer to the Minister; in Queensland it's about 30' (personal communication: Ian Stubbs, Port Hedland solicitor, 1982). Centralised controls over northern development from Perth were seen by some local observers as preventing diversified growth in the Pilbara. State public servants and financial managers in the south were accused of 'never having been north of Geraldton' and of not insisting on the provision of adequate commercial services in towns like Newman. In the latter case in spite of excellent company-provided services there were only seven shops and no new or used car dealership for a population of almost 6000. This is said, for example, to have opened the way for one mining family to make $100 000 a year from selling Holden parts (personal communication: Stubbs).

There was also a strongly held conviction that restrictions over the sale and development of Crown land in the Pilbara on a freehold basis (under provisions of the 1893 *Land Act*) had prevented normal entrepreneurial involvement by the private sector (personal communication: Michael Howieson, Assistant Shire Clerk, E. Pilbara Shire, 1982). Less than five per cent is freehold, for example, the rest being held under pastoral or mining leases of one kind or another. Such circumstances were to provide fertile ground for changes to occur in the administration of the mining towns, even though they were seen by some sections of the union movement as a deal between the companies and the state government.

Both the new housing scheme and the intent of the normalisation programme in Newman received plenty of publicity and political debate in Perth in the two-and-a-half years before transfer took place on 1 July 1981. There is justifiable criticism, however, that there was little open discussion of the matter in the Pilbara itself, and the reasons for this are probably rooted in the paternalistic and authoritarian nature of these mining communities. The attitude exists among some companies in the industry that it is asking for trouble to encourage debate over issues which could end up by promoting a strike. It had been pointed out in state parliament, for example, that in the 1977-78 financial year the workforce in the Pilbara, which made up only five per cent of the state's total, was responsible for 72 per cent of the working time lost in Western Australia (Government of Western Australia, 1979:1328).

The company wished to minimise the opportunities for industrial disputation to occur and an obvious opportunity presented itself in a withdrawal from domestic management in Newman. It was also a change in policy which could be supported by general notions of 'free enterprise' and 'democracy' (Figure 2.1), although the advantages of the latter are very rarely spelt out. In fact Ruth Atkins (1976:51) a political scientist, had said at a seminar on remote communities held at Kambalda in 1973 (following Langrod, 1953) that 'there is no *necessary* theoretical or historical connection between traditional local municipal government and "democracy"'. There is also no guarantee that there will be fewer strikes after the company withdrawal from the town, which is a prime motive suggested by some observers (Thompson, 1981). Indeed, the new employee home ownership scheme (Mt Newman Mining Ltd, 1981) had been in operation for three years in Newman when the town suffered the longest strike in its history.

It appears that the shires had no option but to agree to the normalisation changeover on the best terms they could obtain from the

Figure 2.1 Company town 'normalisation' concepts of the Mt Newman Mining Company, May 1983

The basic concepts pivotal to the normalisation process are as follows:

1 Normalisation is the process whereby a community governed by a non-elected body is transformed into a free enterprise based permanent community governed by a democratic structure consistent with other such structures elsewhere in Australia.

2 The normalisation process does not continue indefinitely as, by definition, it is a transitionary process which terminates when the transformation referred to in (1) above is achieved.

3 Mt Newman Mining Company's original scope of work included (in addition to its mining activities) the construction and administration of the town of Newman, which embraced residential, commercial, civic and utilities services. Policy decisions were subsequently taken by the Joint Venturers which required the Management Company to retreat from its rights and obligations in matters affecting the government of the town of Newman. (These major policy decisions were embodied in the *Iron Ore (Mount Newman) Agreement Amendment Act* and the Proposals of 1979).

4 There is no intrinsic reason why the towns of Port Hedland and Newman cannot, in the future, develop in a similar fashion and by similar means to other comparable permanent towns elsewhere in Australia.

5 Mt Newman Mining Company is now obliged, through the period of normalisation, to rigorously review all its administrative procedures and its involvement in non-iron-ore related activities and to encourage the achievement of 'normality' in the two towns which service the Company's iron ore activities.

Source: Contained in a letter from the Administration Manager, Mt Newman Mining Company to the Chairman, Jabiru Town Development Authority, dated 20 May 1983

companies. There were few short-term advantages from a shire point of view and considerable concern that they would not be able to pay their own workforce a competitive wage in the face of much better conditions in the mining industry. In the longer term there was the likelihood of increased financial viability through the unencumbered acquisition of towns like Newman and a larger rateable base but even here, as we shall see, there were grounds for apprehension. In the words of the East Pilbara Shire Assistant Town Clerk in 1982: 'there is a major difference between remote mining towns where the population, wherever it comes from, is indentured labour; and the situation in small pastoral communities where the people are there because they want to be there—and they'll retire there' (personal communication: Howieson).

Canadian experience suggests that the companies stand to gain most through the creation of a new generation of home owners, and it probably benefits their objectives by reducing labour turnover because of housing commitments. This, in turn, reduces the costs of reproducing labour power for industrial capital (Bradbury, 1983:8). The Mt Newman Mining Company home ownership scheme which started in

March 1981 was based on preliminary research findings suggesting that a very high percentage of employees would purchase a house in the Pilbara. Although those who were most supportive were already the most stable element in the workforce (marrieds with children from a rural area), there was considerable optimism that others who joined the scheme would be less likely to leave the company as a result (MSJ Keys Young Planners Ltd, 1978). Such evidence has led increasingly to a positive discrimination in many mining towns towards a higher percentage of married to single employees, with the situation being reached at Mt Newman Mining where no unskilled singles were being recruited at all (personal communication: Peter Laver, General Manager, 1982).

The normalisation agreement at Newman and the radical critique

The agreement itself finally obtained state government approval in May 1981 and transferred, at peppercorn consideration, responsibility for a range of administrative functions to the East Pilbara Shire Council. Among the most important were road and street maintenance, curbing and drainage, rubbish collection, the community hall, public swimming pool, parks and playing fields. The Shire Council, although located in Marble Bar some hundreds of kilometres to the north, already had responsibility prior to normalisation for things such as the public library, health and building regulations, but it did not exercise any statutory controls regarding town planning and other by-laws until the handover took place (personal communication: John Read, former Shire Clerk, East Pilbara Shire, 1983).

The company provided a branch office for the shire in Newman and a considerable amount of physical plant, as well as 18 furnished houses for shire employees at a nominal cost of $100 per dwelling. But, in spite of this generous support at the outset, there was considerable doubt in the shire about the long-term financial position. An important part of the original iron ore agreements of the 1960s was the amendment of the West Australian *Local Government Act* to include a Section 533B which restricted the rateability of the companies by placing a greatly reduced value on their industrial land. The companies struck a deal with the state government and the shires to overcome the substantial shortfall in rateable incomes rather than change the rating basis on their land. This deal involved a dual system of 'supplementary' and 'transitional' rates amounting in the early years to an annual package for the East Pilbara Shire of about $1 m (*The Australian* 11 April 1983). From the 1983/84 financial year the transitional rate would progressively reduce to zero by 1995, the hope being that an influx of private enterprises to the town would raise normal rateable income by an equivalent amount.

However, the first couple of years after normalisation in Newman were not particularly auspicious owing to the effects of recession and a depressed export market for iron ore. All the Pilbara Shire Councils took the view that the only sensible long-term basis for the administration of the company towns would be to allow normal rateability to apply to company owned land and railroads and they were hopeful of achieving this at some time in the future (personal communication: John Read, former Shire Clerk, East Pilbara Shire Council, 1983). It is interesting to note for the record that one way new industries and residents were encouraged to move to Newman was to make it mandatory for the company to supply 20 per cent of building blocks in new residential sub-divisions to the government for public auction. When demand from the private sector outstrips this provision it is likely that the shire itself would consider acting as the land developer in order to promote growth in the town.

The situation which had developed in Newman is not unique in Australian company town history and were the East Pilbara Shire councillors to have examined the growth of Whyalla, the steelworks town and port in South Australia, they would have had even greater grounds for pessimism. In that instance the Broken Hill Proprietary Company (BHP) had received several important concessions from the state government under the *Steelworks Indenture Act* of 1958 to encourage them to build a steelworks in the town. Roy Kriegler (1980:7–8) has shown that not only did the terms of the Act force the Whyalla Council to locate its own industrial area far from the town (because the land in between was controlled by BHP), it also resulted in an unsatisfactory basis for council income:

> As an additional benefit for signing the 1958 Indenture, BHP is not required to pay any rates or taxes on its vast industrial sites in the north and south of the town. A senior officer of the City Council described this as a 'scandalous loss of revenue of which Whyalla had been robbed'. He explained how BHP makes an *ex gratia* payment of between $15,000 and $20,000 to the Council in lieu of rates, which represents only about two per cent of the tax that they would have to pay had the land been rateable.

Similar findings by Aungles and Szelenyi (1980), underline the impotence of local councils which have no financial muscle, and serve to emphasise that the advent of local government to a company town does little in itself to change the economic realities for single industry dependent communities. The achievement of 'normality' under these circumstances does not extend much beyond the introduction of local government institutions and certainly does not threaten the underlying powers of the companies themselves. Here too, is something to be borne in mind during our interpretation of the events which were to occur in Jabiru.

It is necessary, in concluding this section, to make some assessment of the radical critique of government and company policy in the Pilbara towns. There is a strong case to suggest that the normalisation process is functional to the long term interests of mining capital and few would dispute this but, at the same time, there was little opposition to the changes either from within the towns or from the main political parties in Perth. The radical critique hinges on the long-term disadvantage to residents in the greater living costs associated with housing ownership and the progressive loss of company subsidy of town amenity, yet the alternative, a continuance of company control, has found no support in the radical literature in the past. It seems therefore that the thrust of the political economy critique of these events has more to do with overall problems of mining employment in capitalist society than it has with the circumstances of the normalised towns themselves (Lea and Zehner, 1985).

There may well be alternatives to an Australian mining industry based on the dominance by multinational companies but it is very unlikely that the nationalisation this would require will take place in the foreseeable future. Even if it does occur, it is hard to judge the manner in which it will benefit the residents of the remote Pilbara towns. A perennial risk for some radical scholars is that they may become stuck in what has been termed elsewhere as the 'structural trap of informed inactivity'. In other words, opposition on theoretical grounds is made to changes which may ameliorate local conditions and attract working-class support because those changes may also serve to support the overall capitalist system.

Concluding comments

A common factor running through this brief review of local government in new Australian mining towns is the importance of private property ownership. It can be seen in the earliest examples as a concept which suited the purposes of the Victorian state government anxious to create a steady source of income through the sale of Crown land, 'Local government appeared as a means of subdividing Victoria into marketable pieces of real estate and of carrying out subsequent property improvements' (Barrett, 1979:136). In addition, the ownership of property at an individual level was seen as a prime means of social mobility and 'many decisions affecting a community could be made privately by any owner of local property' (Barrett, 1979:15).

Seen in these terms, the initial absence of private property rights for residents in the new Pilbara company towns may be judged, in retrospect, to have been an aberration. The history of Australian local government is full of references to property ownership and it seems that it is difficult, if not impossible, for this bottom tier of governance to flourish under conditions which do not actively support it. 'It is fair to conclude that their activities have not seriously challenged the forces of privatism in the cities. Indeed, Australian local governments seem to embody a property-related sense of mission that interprets the "public interest" in market related terms' (Parkin, 1980:378). This is a realisation which has profound implications for the future evolution of government in Jabiru.

Ownership of property is also closely connected with the nature of local politics in Australian towns and cities and it is difficult to see how local political activity can flourish in situations where private property ownership is absent or, as in the normalised Pilbara towns, significantly restricted:

> The determination of property owners and those who use land for profitable purposes to control access to councils through restricting the franchise reflects what local politics is all about: it has as its focus control over the use of land. Local politics is primarily over territory: who owns it, how much is it taxed and how services are provided to it (Chapman and Wood, 1984:48).

In the Victorian examples, the central importance of private property is expressed in terms of furthering social mobility through opportunities for personal capital accumulation. 'A prudent man might invest his savings in buildings or land and then wait for the expected increase in property values resulted from community processes, the benefit accrued not to the community but to the private title holder' (Barrett, 1979:15). In the Pilbara, the goal of community stability advanced by state government and companies alike rests on the assumption that the determining influences of property ownership will act as a sort of social 'multiplier', supporting the various other characteristics of Australian local government and town life to which we have referred.

There are good grounds for doubting whether such assumptions hold for remote single industry communities where many of the traditional opportunities for social mobility through house and property ownership are constrained. This is the case in the modern Australian examples we have identified and it has a similar constraining effect on local politics. Popular appeals by the West Australian government and the iron ore companies to recreate 'normal' conditions of the type which emerged unaided in the Victorian gold towns a century ago neglect this reality and some of its consequent effects on the future social structure of these communities.

Part II NEW TOWN IN THE URANIUM PROVINCE

3 Uranium discoveries and development choices

> After consideration of all factors, we propose a solution which, if a decision is made that uranium mining is to proceed, provides a reasonably satisfactory accommodation between competing interests and the conflicting uses to which land in the region can be put. This is subject to one qualification. The principle threat to the welfare of the Aboriginal people, and the one they most fear, is constituted by the large numbers of people who can be expected to enter the area. We make a number of recommendations designed to minimise this risk; in particular we recommend strict limitation on the size of the town and the use of the area by tourists (Fox et al., 1977:9).

Government interest in the development of the Alligator Rivers Region is very recent in comparison to the Aboriginal presence in the area which is thought to date back for at least 20 000 years (Jones, 1984), and European contact which reaches only to the mid-nineteenth century. The first official plans for a National Park came in the mid-1960s but is was not until early in the following decade, after the Ranger uranium discoveries in October 1969, that serious moves were begun to modernise the regional infrastructure with new roads, services and a town. Several competing interest groups with a stake in the Region's future development were brought together in the Ranger Uranium Environmental Inquiry held between 1975 and 1977. It was a landmark in Australian environmental planning, as we have noted, and provided a platform to air many of the basic conflicts in modern resources development: between the 'new' rural industries of open-cut mining and tourism and the 'old' pastoral and semi-subsistence Aboriginal economies; between a new southern-based environmentalism and the older and conservative developmental priorities of the north; between newly won Aboriginal rights to land ownership and the economic interests of other rural activities; between the political and economic hegemony over development held by the southern states and the burgeoning spirit of regional identity in the north, and so on. In this chapter we examine the nature of some of these tensions as they have affected developments in the Uranium Province together with the

33

compromises and choices which were made at the time of the Ranger Inquiry.

The primary focus here is on one important aspect of the development debate, the creation of a new regional town to accommodate the expected influx of people drawn by the mining and tourist industries. The eventual outcome, as indicated in the passage from the Ranger Inquiry Report quoted above, was to be a 'closed' town rather than a diverse regional centre. Some of the development conflicts which were present at the time of this decision remain unresolved today and one of our intentions is to clarify them by tracing their origins to the period before construction of the town began.

A brief history of developments in the Alligator Rivers Region

Poor surface communications in the wet season and a lack of obvious economic prospects other than a precarious pastoral industry based on the feral buffalo combined to restrict a European presence in the area to minimal proportions historically. The coast is known to have been frequented by Macassan fishermen since the seventeenth century (Macknight, 1976) and was surveyed by the expeditions of King, Wickham and Stokes in the early nineteenth century. There were short-lived British military settlements at Fort Dundas (Melville Island), Fort Wellington (Raffles Bay) and Port Essington between 1824 and 1849 but none was successful, giving the Region a poor reputation with the builders of empire.

The cattle industry enjoyed a short life in the 1880s and 1890s before suffering a demise through the combined effects of poor communications and tick-borne Redwater Disease (Ranger Inquiry Evidence: 1517). Only the buffalo industry, said to have been started by Paddy Cahill at Oenpelli, survived, and exists today with a base at Mudginberri pastoral lease. A colourful account of buffalo shooting in this area in the 1920s is found in the book by Warburton (1934), a returned soldier who won a shooting concession there after World War I.

Before describing the changes which accompanied the uranium discoveries, it is necessary to define the Region in rather more detail than we have attempted hitherto. The catchments of the East, South and West Alligator Rivers form an areal unit which covers most of the present important land use activities (Map 3). The first uranium discovery worth commercial exploitation came in 1953–54 in the headwaters of the South Alligator River. Jack Smith, a prospector working for United Uranium NL, one of many small exploration companies active in the area at that time (see *The Australian Financial Review*

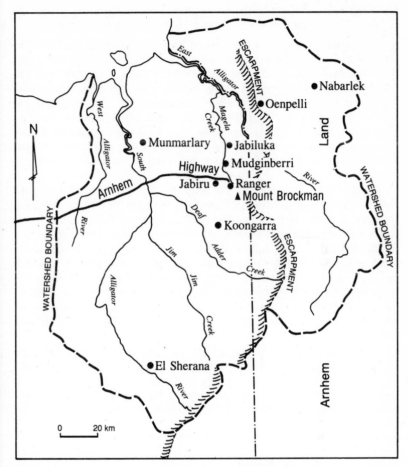

Map 3 *The Alligator Rivers Region, Northern Territory.*

18 February and 27 May 1954), is credited with finding the rich El Sherana strike:

> For two weeks they combed the craggy bluffs without finding a trace of anything. Then Jack made the strike of the year. He had been prospecting all day in high, steep gullies, and had arranged to meet his mate after dark at the mouth of Stag Creek, a tributary of the South Alligator. Just on dark, Jack reached the summit of an almost perpendicular ridge to find his geiger beginning to tick wildly. He had only been using a geiger for about a month, but there was no mistaking the crazy tattoo in the earphones...
> They named the lease after the three daughters of...one of the prospectors

in the team...By mixing the names Ellen, Sharon and Anna, they got the magical-sounding title of 'E1 Sherana' (Annabell, 1971:115–16).

Mining at E1 Sherana had ceased by the mid-1960s. In 1965 The Northern Territory Reserves Board made a proposal for a 6410 square km National Park in the area between the South Alligator and the Arnhem Land Aboriginal Reserve. Neither this, nor a commonwealth study made in the previous year (Commonwealth of Australia, 1964) which looked at the upgrading of the road from Darwin, resulted in any changes in the Region itself. The Park was not to be finally gazetted for another ten years and the transport study came to the conclusion that the pastoral and buffalo industries were too small to justify any improvements to the road.

Further proposals for the National Park came in 1967 and 1968 before ministerial approval of land reservation for the purpose was secured in 1969. But, as Saddler (1980) observes, full pastoral leases were also approved in the same month for the two buffalo properties of Munmarlary and Mudginberri (they had been grazing licenses in former times): 'We see here a clear example of the attitude towards national parks which has prevailed ever since the establishment of the first national park in Australia, the Royal National Park of Sydney: only if a tract of land is demonstrably useless for any other purpose should it be considered for inclusion in a national park' (Saddler, 1980:189).

In the meantime, four major uranium strikes were to be the catalyst for extensive public and private activity in the Region before the Kakadu National Park was finally declared in 1975. Figure 3.1 contains a chronology of significant events affecting the Region from the time of the uranium discoveries until the Ranger Inquiry hearings six years later.

The discovery of the Ranger orebodies

Late in 1969 came the first suggestion that there were large deposits of uranium in the Geopeko prospecting area which were far more important than the earlier finds in the South Alligator. Renewed interest in the Region by mining groups had followed the publication in 1968 of a new 1:500 000 geological map of the Katherine-Darwin Province (Ryan, 1972). This showed that rock outcrops near the Arnhem Land escarpment were much older than previously thought (as a result of age-dating) and were closely analogous to the uranium-bearing formations found to the west at Rum Jungle.

Geopeko made arrangements for an airborne search using magnetic and radiometric techniques, but equipment problems and the

Figure 3.1 Chronology of events leading to the Ranger uranium discoveries and the Ranger Uranium Environmental Inquiry 1975-77

July 1967	Gondwana Joint Venture formed by Peko-Wallsend Operations Ltd and Electrolytic Zinc Co. of Australasia Ltd to explore for uranium and base metals in northern Australia. Geopeko Ltd appointed to conduct exploration
Nov. 1968	Weems Report on proposal for National Park
Mid-1969	Noranda (Australia) Ltd detects anomaly in Ranger area
June	Minister for the Interior approves establishment of National Park in principle
October	Geopeko ground party confirms anomalies and pegs leases
Feb. 1970	Georgetown exploration camp established
May	Nabarlek uranium deposits discovered by Queensland Mines Ltd
August	Koongarra orebody discovered by Noranda
Late 1970	Jabiru exploration camp and airstrip completed
Nov. 1971	Jabiluka 1 orebody discovered by Pancontinental and Getty Oil Joint Venture
December	Peko and EZ enter into Ranger Joint Venture Agreement
Feb. 1972	Dept of National Development commissions feasibility study for a new regional centre
	Negotiations start between Joint Venturers and commonwealth for a Special Mineral Lease
	Uranium sales contracts concluded with two Japanese power utilities
March 1973	New government in Canberra defers Ranger mining lease and seeks to appropriate Northern Territory uranium through the *Atomic Energy Act*
April	Commonwealth authorises detailed town planning studies for proposed regional centre
June	Cities Commission Workshop in Darwin
September	Jabiluka 2 orebody discovered
December	Commonwealth suspends exploration rights in area proposed for National Park
April 1974	Second Report of Aboriginal Land Rights Commission
October	Ranger Venturers and commonwealth enter into the Lodge Agreement to permit the project to proceed and honour Japanese sales contracts
March 1975	*National Parks and Wildlife Conservation Act* passed
July	Ranger Inquiry set up under the provisions of the *Environment Protection (Impact of Proposals) Act, 1974*
October	Terms of the Lodge Agreement elaborated into a Memorandum of Understanding
Oct. 1976	First Report of the Ranger Inquiry
December	*Aboriginal Land Rights (NT) Act* passed

arrival of the wet weather delayed this until mid-1970. In the meantime Noranda, the company which held the authority to prospect in the adjoining area, detected high radiation in the Geopeko prospects while searching their own area from the air. This information was passed on to the Ranger Joint Venture (Geopeko was their exploration company) which sent a ground party to investigate and peg two leases in October 1969.

Materials for a small camp were left at the site and early the following year, when the worst of the wet season was over, four men were ferried in by helicopter. Hand-held scintillometers (which measure radiation against the general background radiation) showed values of 'up to 30 times background' (Ryan, 1972:298). A further airborne search soon revealed five more anomalies over a distance of some 6 km (Map 4). Danielson (1984) should be consulted for a more recent review of the mineral potential of the whole Ranger Project area.

The existence of an anomaly does not necessarily mean that uranium is present in commercial quantities as it is an unstable element which passes through many stages before ending up as lead. Thus it is necessary to follow aerial exploration by soil sampling, trench cutting (consteaning) and drilling. *The Australian Financial Review* (19 July 1973) reported that because only two metres of sand cover is sufficient to mask an anomaly, the explorers also employed various ingenious techniques to locate them. They found, for example, that the pandanus palm which is common in the area concentrates uranium, as its roots often penetrate below the sand cover. The same is true of termite hills and both Queensland Mines and Noranda are reported to have made discoveries in this way.

By now it was clear that Ranger Anomalies 1, 2, and 3 represented a discovery of world significance and that major site work was necessary to delineate them. The original exploration camp was called Georgetown and located beside a billabong of the same name close to Ranger 1 (Map 5). The story behind the name is that Geopeko, the exploration company, used to occasionally receive mail addressed to 'George Peko' and personnel soon took to calling the company 'George', leading to the naming of the camp (*NT News* 15 December 1979). It proved to be a poor location however, with insufficient room for expansion and subject to flooding. A new camp and airstrip called Jabiru was built some four km away where a good water supply was found. The story behind this name is that Sir John Proud, then Chairman of Geopeko, saw a large bird when he was visiting the site and on being told it was a jabiru suggested its name would be suitable for the camp (*NT News* 15 December 1979).

The exploration team was well aware that their find lay within the boundaries of the proposed National Park and realised it was in the Joint Venture's interests to demonstrate the size and potential of the discovery as soon as possible (Ryan, 1972:298). The new camp buildings, airstrip and roads were completed within three months with most of the material plus six months' supplies being brought in from the southern states. Many of these original buildings remain today in the area of temporary housing known as Jabiru East.

According to Rob Ryan, the managing geologist, contact with local Aborigines first occurred as ground crews moved onto the site in February 1970:

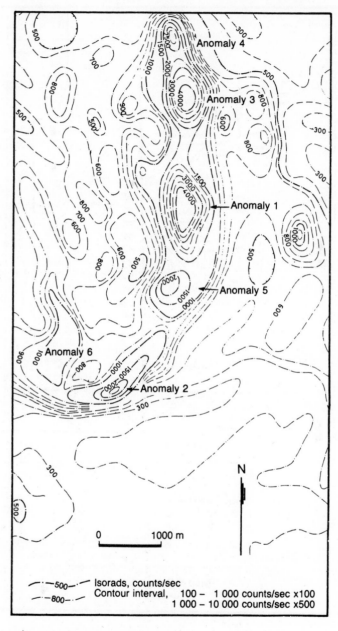

Map 4 *The Ranger anomalies.*

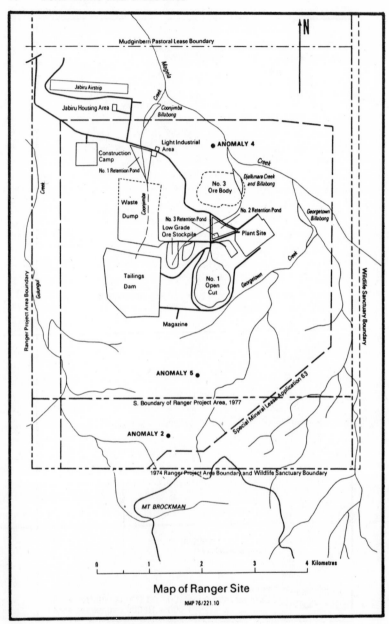

Map 5   *The Ranger Site.*

I should say that we were very fortunate in being apprised—very early in the piece, of the aboriginal interest in the area by Mr Peter Balmanidbal, who saw us on the day we were moving our first survey crews into the area; indicated that the escarpment was of particular interest, asked us where we would be working and we said we would be out on the plains and we were not interested in the escarpment, and that was the conclusion of the conversation (Ranger Inquiry Evidence, 3 October 1975:1449).

Two important considerations arose from this early experience: the interaction of new mining activity with local Aboriginal society; and the increasing level of conflict between geological exploration and the natural environment. The preservation of written evidence about these twin concerns is an important by-product of the Ranger Inquiry which was soon to follow, and forms a valuable case study of the exploration activity which was occurring in northern Australia at the time.

As far as environmental impact was concerned the experience may have been atypical, as there is extensive evidence to suggest that the Ranger Joint Venture took a great deal more care over the environmental effects of their operations than was the norm at this time (see Ranger Inquiry Evidence, 3 October 1975:1446–64; and *NT Parliamentary Record* Part 1, 28 November 1978:513). Some of their actions regarding the Aboriginal presence in the area and the proximity of the Ranger anomalies to the sacred sites on Mount Brockman are rather more questionable. Counsel for the Inquiry was particularly interested in the way the southern boundary of the mining area was moved in order to accommodate Ranger Anomaly 2. It can be seen from Map 5 that this line runs extremely close to the foot of the escarpment and the sacred sites of Djidbidjidbi and Dadbe and might be expected to cause considerable concern to the local people. Mr Cummins, the Ranger Inquiry counsel, engaged in lengthy questioning of Mr McIntosh, a senior company representative. We reproduce part of it here:

Mr Cummins: Well might I put this directly to you. Do you remember reading this report about that time: 'they'—referring to the Aboriginals—'sincerely believe that the transgression of the borderline entailing the splitting of rocks and the vibrations, caused by drilling rigs and earth-moving equipment in action would bring on an unimaginable disaster, that is the collapse of the Mt Brockman cliffs, and with it, the perishing of the people over a wide area'.
Mr McIntosh: I've read that (Ranger Inquiry Evidence, 8 October 1975:1566).

Later, the questioning turned to the company's negotiations with the Aboriginal custodian of the sacred sites about moving the boundary closer to the escarpment:

Mr Cummins: I see, and then—I'm suggesting to you that Peter

[Balmanidbal] showed some concern, and commented that you could work as much as you liked—indicating over there and pointing to the east and west. And then, of course, you indicated, did you not, to this meeting, that really you wanted to work south of the line. Is that right? And he said that that was alright?
Mr McIntosh: Down to a point—up to a point.
Mr Cummins: Yes.
Mr McIntosh: To a particular tree.
Mr Cummins: And then—that the boundary was then changed to 70 metres south of the original line?
Mr McIntosh: That's about it. Yeah.
Mr Cummins: And it was in those circumstances that Peter was given $2.00 and a packet of cigarettes—and that set a pattern, did it not, for the payment...(tape inaudible)...Well of course you remember, do you not, that you made reports to the Company about these matters. Did you?
Mr McIntosh: I beg your pardon?
Mr Cummins: You made reports to the Company.
Mr McIntosh: I did, I did, yes (Ranger Inquiry Evidence, 8 October 1975:1568).

Although he does not come out of this segment of the Inquiry evidence in a positive light, Alan McIntosh demonstrated in other places his keen and sympathetic interest in Aboriginal welfare and environmental issues. He was later to become Services Manager at the Ranger mine and occupied this position until he retired in 1983. The southern boundary was moved further south, twice, with the agreement of Peter Balmanidbal, ending up some 200 metres closer to the sacred sites and became the line used by the Company in their application for a Special Mineral Lease in 1974. The Ranger Inquiry Commissioners expressed concern about its proximity to the escarpment and recommended that the southern boundary be returned to its original position further north (Map 5). The determination was accepted by the commonwealth and meant that most of Ranger 2 now lay inside the National Park and would not be mined (Commonwealth of Australia, 1977).

This illustration has been recounted in some detail as an early example of the emerging conflict between the search for minerals and preservation of sacred sites which was to dominate exploration activity for the next decade. Succeeding discoveries at Nabarlek, Koongarra and Jabiluka heightened pressure on the commonwealth to consider the future of the Uranium Province as a whole leading directly to the Ranger Inquiry itself. Before it took place, however, the commonwealth investigated some of the requirements which major mineral developments in the Region would demand, such as the upgrading of the main road to Darwin and the possibility of a new town. The Parliamentary Committee on Public Works decided in 1972 to improve the road to all-weather standards and bridge the South Alligator, resulting in a great increase in visitor traffic and serious damage to Aboriginal sites and the natural environment (Saddler, 1980). Plans for a regional centre to act as a hub for mining, tourism and pastoral activities were well advanced by the time the Inquiry was announced in 1975 and supported by a strong pro-development lobby determined to maximise the opportunities provided by the uranium discoveries.

Plans for a regional centre in the Uranium Province

Conservationists were among the first to realise that the mines were not the only threat to an area under consideration for a National Park and that several townships might be built. The Darwin Conservation Society is reported in 1971 to have sent 'sternly worded, 10 page reports to the Minister for the Interior and the Administrator of the NT ...listing the ecological dangers of mining, and suggesting methods by which they could be minimised'. Chief among the suggestions was that only one town and processing plant should serve the entire field and that 'such a town be limited to a pre-determined size, and be surrounded by a belt of national park, as should each mine and treatment plant' (*Canberra Times* 2 November 1971).

Plans were needed urgently to contain the impact of mining and tourism activity, leading the Department of National Development to commission a study of future urban requirements for the Region in February 1972. Subsequent reports from this study (Cameron McNamara and Partners et al., 1972a; 1972b) recommended building one new regional centre on a central site approximating to the present location of the town of Jabiru. It would be designed to serve the needs of all new economic activities within 100 km and was based on the interesting premise that the government would have to pay for the initial costs of about $27 m, because 'the separate mining companies ...would not possess sufficient incentive to combine their working communities within a single urban area' (A. A. Heath and Partners et al., 1973:9). This had certainly been the case in the Pilbara and in Queensland's Bowen Basin, where considerable duplication exists among small mining settlements (Zehner and Lea, 1983).

The site itself was chosen from four possibilities (Map 6) arising from a sieving process based on the identification of six restraining factors: Aboriginal Reserves; the escarpment; the proposed National Park (and pastoral leases); prospective mining areas; and the location of anomalies and mineral leases (Simpson, 1980). Only one location, Site B, was shown as having no major restraints of this kind and offered an area of some 2000 ha, 8 km west of the Ranger exploration

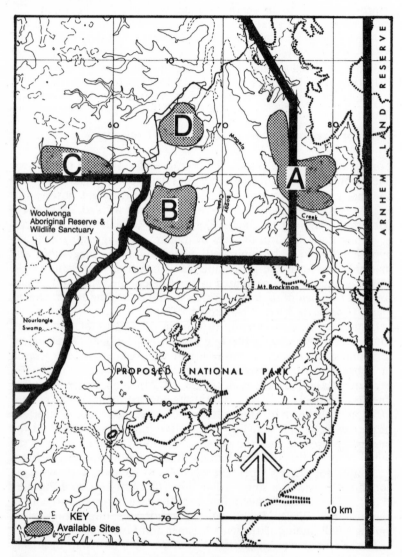

Map 6   *Possible locations for the town of Jabiru.*

camp (Map 7). Important assumptions were that half the housing requirements of the companies would be provided by the Northern Territory Housing Commission with the remainder coming from the Commission using company funds. Additional housing for the public

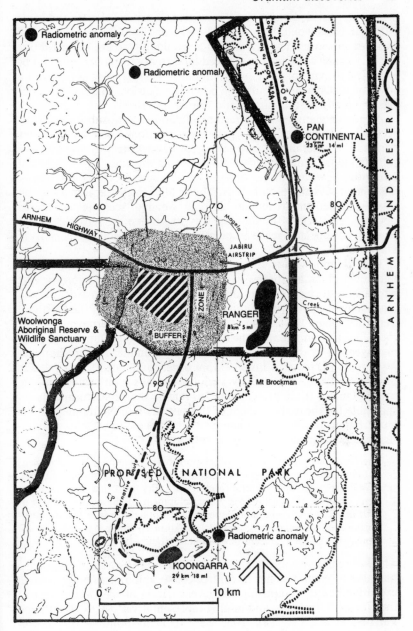

Map 7   *The site for a new town.*

would be provided under the Commission's general charter and some $19 m in costs would be recovered by leasing land for residential and business purposes. The private sector would thus secure leasehold title based on reserve prices for serviced land (rent plus premiums) and be expected to pay for their own buildings. The planners anticipated a town of 10 000 would result with possible extensions to double this size based equally on the mining and tourist industries.

Although many of these recommendations were subsequently rejected by the Ranger Inquiry, they are mentioned here because they formed the generally accepted basis for development by government, mining industry and commercial interests at the time. They paid scant attention, however, to the views of local Aborigines or conservationists and it was these omissions, above all, which the Ranger Commissioners set out to redress. But, as we shall see, the 'unfettered development' school of thought was to survive the setbacks of the Ranger Inquiry and received plenty of official encouragement from Darwin.

The commonwealth authorised detailed site surveys, engineering and town planning studies in April 1973 which led to more reports which were to be highly influential in the eventual construction and management of Jabiru. At this stage the Department of the Northern Territory sought the assistance of a new Whitlam government agency, the Cities Commission, in the appointment of consultants and in the management of the town development programme. A workshop was organised in Darwin in June 1973, in order to initiate the town design process and was attended by various specialists on remote mining towns. One of those present was Jos Agius, the person primarily responsible for the earlier design of the new town of Nhulunbuy on the Gove Peninsula. He was adamant that the new town must be constructed by a single development authority:

> The Government has to take full responsibility for establishing a nucleus of town centre facilities, provide essential goods and community services, as well as establishing a central funding authority through a single management structure responsible for the development of the town in terms of both physical and community enterprise. In other words, a Development Corporation designated by formal legislation enabling it to take the responsibility of total concept development (Agius, 1973:1–2).

The Jabiru design consultants took this advice to heart and proposed a manager or commissioner to control construction, assisted by a small executive board 'comprising top level officers of the Department of the Northern Territory and mining representatives' (A. A. Heath and Partners et al., 1973:84). They were to be responsible for coordinating all public and private sector development and would retain government control. Nothing was said at this stage about eventual community participation in managing Jabiru but there is information from elsewhere to suggest that some form of local government was the ultimate intention (Ranger Inquiry evidence by T. Brooks, Local Government and Services Branch, Department of the Northern Territory, 28 May 1976).

A structure plan was also proposed in the first stage of the design study (Map 8) as well as the incorporation of a large recreational lake. Many features were introduced which were to form the basis of the final plan five years later, such as road and housing layouts, pedestrian walkways and the location of the town centre. Towards the end of 1973, the commonwealth halted further planning work on the regional centre pending decisions about the whole question of uranium mining in the Northern Territory and the creation of the Kakadu National Park. In addition, the Woodward Aboriginal Land Rights Commission had already brought down reports in 1973 and 1974 recognising the legitimacy of claims over unalienated Crown land by Aborigines and their rights to veto mining activity on it (Woodward, 1973; 1974). Indeed, the Northern Aboriginal Land Committee Inc., the forerunner to the Northern Land Council (NLC), was opposed to mining on the Ranger site at this time (Fox et al., 1977).

In July 1974, Ranger Uranium Mines applied for a Special Mineral Lease of some 2869 ha to enable them to commence mining (Map 5). Later in the same year the two Ranger partners signed a memorandum (the Lodge Agreement) with the commonwealth to allow them to mine in partnership with the Atomic Energy Commission (AAEC). This came about because the Labor government wished to be involved in the industry having failed to take possession of all uranium in the Territory the previous year. Ranger would still be responsible for milling and delivering the uranium but the AAEC would come in as a partner providing 72.5 per cent of mine, mill and infrastructure costs. The details were elaborated and formalised into a Memorandum of Understanding in October 1975.

During 1974 the commonwealth had passed the *Environment Protection (Impact of Proposals) Act* which provided a suitable means of establishing a far reaching inquiry about the impacts of the uranium discoveries. Three Commissioners, a senior Judge of the ACT Supreme Court, a civil engineer and a Professor of Preventative and Social Medicine, were appointed in July 1975. The resulting Inquiry was to have the effect of freezing further development in the Region for the next three years.

Vested interests in the Uranium Province

It is important to identify the groups at the Ranger Inquiry who were represented as having a primary interest in the future of the Region,

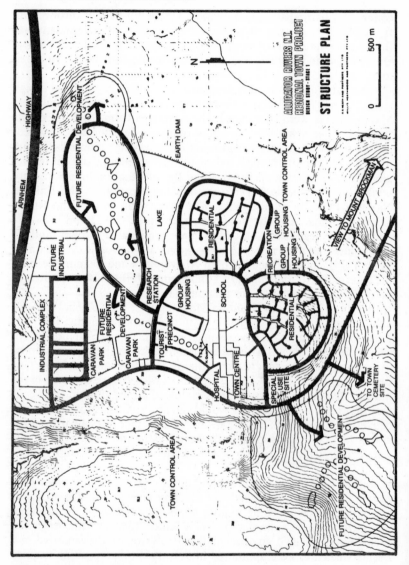

Map 8   *Jabiru Structure Plan, 1973.*

before turning to the guidelines established for the creation of Jabiru. Six groups, together with the commonwealth, were formally represented by legal counsel (Figure 3.2). Apart from the government, these bodies are broadly divided on the basis of their attitudes to uranium development. In favour, besides the mining companies and the AAEC, were several commercial hotel and trading enterprises; against were the pastoral company, environmental groups and Aboriginal organisations. Two of the mining groups, Dampier and GEMCO, were not in the uranium industry but were BHP-owned companies with extensive exploration interests in the 'Top End'.

On the other side, Northern Pastoral Services held pastoral leases over Munmarlary and Mudginberri stations and processed domesticated and feral buffalo meat through the new abbatoir at Mudginberri. Evidence to the Inquiry suggested a rather precarious financial position and a wish to sell the company should mining proceed (Fox et al., 1977). Aboriginal and environmental organisations were also against mining, seeing it as incompatible with Aboriginal culture and heritage and the area's many outstanding natural

Figure 3.2 Organisations represented by legal counsel at the Ranger Uranium Environmental Inquiry 1975-77

Government
    Commonwealth of Australia
Statutory Authority
    Australian Atomic Energy Commission
Mining Companies and Organisations
    Ranger Uranium Mines Pty Ltd
    Pancontinental Mining Ltd
    Noranda Australia Ltd
    Dampier Mining Company Ltd
    Groote Eylandt Mining Company Ltd (GEMCO)
    Australian Uranium Producers Forum
Pastoralist
    Northern Pastoral Services Ltd
Aboriginal Organisations
    Northern Land Council
    Oenpelli Council
Environmentalists
    Australian Conservation Foundation
    Friends of the Earth
    Conservation Council of South Australia
Commercial Enterprises
    Optiz 'Cooinda' Enterprises Pty Ltd
    Roper Bar Trading Pty Ltd (South Alligator Motor Inn)
    Mr & Mrs K. Hill (Border Store, East Alligator)

Source: After Fox et al., 1977, Appendix XI

attributes. The Aboriginal stand was to lead to a temporary impasse after the hearings were finished and before they agreed to allow mining to proceed.

Although the commonwealth's decision in 1977 was in favour of uranium mining, it also approved the recommendation that much of the land in the Region, including the mineral leases, be returned to Aboriginal ownership. This occurred because the Alligator Rivers Stage 1 Land Claim was heard concurrently with the main Inquiry by the Ranger Commissioners under the terms of the *Aboriginal Land Rights (NT) Act* of 1976 (Altman, 1983). It was necessary under the provisions of the Act for the commonwealth and NLC to reach agreement about royalties and other conditions before the Land Council could give consent to an application to mine. Many rounds of negotiations were held between the parties in 1978 before terms and conditions were finally agreed and Aboriginal consent secured (Parsons, 1978). The Ranger Agreement, though not without its critics (see Roberts, 1981), provides for substantial royalties amounting to 4.25 per cent of the value of mineral production (Commonwealth of Australia, 1978) and employment of Aborigines in the mine and in associated service occupations (Woods, 1979; O'Faircheallaigh, 1985).

The Ranger Inquiry is generally held to have been exemplary, despite criticism from some of the interest groups involved (Roberts, 1984; Woods et al., 1978) and its very success in protecting the natural environment may actually prevent governments in the future from pursuing such investigations:

> The inquiry, together with the period subsequently allowed for public debate, delayed the decision to proceed with mining by many months. The firmness and independence of the commission when subjected to pressure inconvenienced the government, and the commission's recommendations placed considerable constraints on the courses of action open when implementing the decision to proceed. No government readily accepts such impositions, particularly when by doing so it is likely to be faced with recommendations which run counter to its basic convictions (Formby, 1981:209).

Guidelines provided by the Inquiry for developing the new town of Jabiru were accepted by the commonwealth in all important respects and set the scene for the creation of a mining enclave rather than the diverse regional centre of previous proposals.

Guidelines for a new town

The tone of the recommendations is captured in the title to chapter 12 of the Inquiry Report 'Accommodation of Mine Workers and Their

Families', and is an early indication that the economic functions of the new town were to be severely restricted. A package of pre-conditions and guidelines were devised to ensure the admirable objective that any new settlement in the Region would be designed to minimise effects on local Aboriginal people and the natural environment. There are limits, of course, to ensuring such an outcome in a free and democratic society and it is likely that the controls imposed over the development of Jabiru represent the boundary of popular toleration. The NT Chief Minister was later to make them the butt of regular attacks in the Legislative Council, saying on one occasion that 'I believe quite frankly, that the director [of the Parks Service] can control almost every action down to the growing of cabbages in the park' (Paul Everingham, *NT Parliamentary Record* Part 1, 10 June 1981:1129). This is not to deny the necessity for special conditions governing the nature of built development, given the peculiar circumstances of the situation, but such controls will rarely curb strong pressures for urban growth indefinitely.

Popular demands for economic diversification and relaxation over development controls have surfaced in Nhulunbuy in recent years, suggesting that only a short period elapses before initial constraints must be reassessed (Lea, 1984). This may have some parallel significance for Jabiru as Nhulunbuy was built as a 'closed' town in 1972 and used as a model in the early planning studies for the regional centre in the Uranium Province.

The Ranger Inquiry guidelines comprised a package of seven items covering land use planning and land ownership pre-conditions, the town site, construction activity, the economic base, town size, behaviour of residents and phasing of mining development. They were all to be accepted by the commonwealth, other than a suggestion to reassess the suitability of an alternative town site and the imposition of planned phasing for the Ranger and Pancontinental prospects.

A crucial determinant of future urban development was the proposal to grant Aboriginal title over all land in the area later to become known as the Kakadu National Park Stage 1, apart from the town site itself and the small motel at Cooinda (Map 9). It paved the way for the establishment of the National Park through a lease-back of the land by the Aboriginal owners to the Parks Service for a 99-year term. The granting of Aboriginal title over the mining prospects enabled negotiations to commence with the companies under the terms of the NT Land Rights legislation of 1976. Of primary importance for the new town was that its development would now fall under the provisions of the *National Parks and Wildlife Conservation Act* (1975) and the plan of management for the Park. This device ensured that the common-

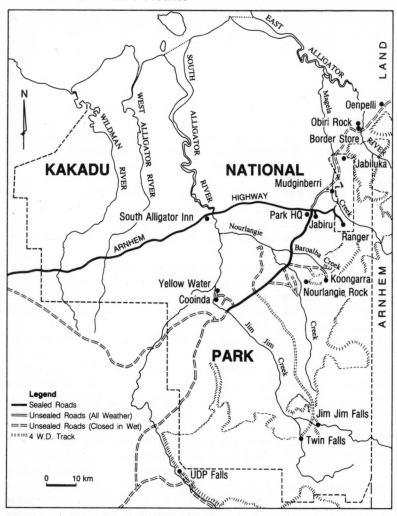

Map 9   *Kakadu National Park.*

wealth would have the final word over major developments in the Region in consultation with Aboriginal groups.

Other important recommendations were in the form of quite detailed guidelines for the physical development of the town, with some of them being qualified to enable a reassessment at a later date. Thus tourists were not allowed to be accommodated in the town *for the time being*, and its size was to be kept to a maximum of 3500 *subject to a periodical review.* This left the door open for the Parks Service to vary the conditions at a later date should circumstances demand it and Aboriginal opinion allow. The location in the National Park gave rise to controls over personal behaviour not before experienced in Australian mining communities. Firm environmental control meant the monitoring of residents and the retention of ultimate authority by the Parks Services, making it difficult to see how normal local government could be introduced at a later stage. When the matter was raised subsequently in the Legislative Assembly, the Chief Minister was moved to declare that the introduction of a rating system in Jabiru would be 'one of the wonders of the western world' (Paul Everingham, *NT Parliamentary Record* Part 1, 24 August 1983:220).

In chapter 1 we indicated that two important questions for our analysis were the reasons why a novel 'development authority' approach was chosen for Jabiru, and the causes underlying the speedy introduction of local government. These and other matters now became the responsibility of a new Territory administration on the attainment of self-government in 1978. Some aspects, such as the idea of a statutory authority and the progressive involvement of residents in town affairs, clearly dated back to the pre-Ranger Inquiry days. The former was strongly canvassed, as we have noted, in the 1973 Cities Commission Workshop and the latter was consistent with the broader trend towards normalising the administration of smaller towns in the Territory. In fact, a senior official in the Local Government and Services Branch of the NT Administration suggested at the Ranger Inquiry that local government would come to Jabiru once the town was 'established on a firm footing' (G. W. Godwin, Ranger Inquiry Evidence, 28 May 1976:3).

All of this was put forward, however, before the new and much more limited development strategy for Jabiru became a reality, and before it was known that the town's future would be determined by the National Parks Service in consultation with the NLC. The new Territory government was thus faced with a situation of considerable uncertainty tinged with relief that the long awaited developments in the Uranium Province were underway at last. Next, we examine the creation of the Development Authority, the three-year construction period from 1979 to 1982 and the way in which residents became progressively involved in town affairs, culminating in the election of the Jabiru Town Council in mid-1984.

4 Jabiru

> One of the biggest projects underway at present is the development of an entire town in the Uranium Province. It's a town called Jabiru. Destined to be the third largest settlement in the Northern Territory... It will be around for generations to come and will stand as one of the most ambitious, carefully planned and implemented developments ever undertaken in Australia ...(*NT News* 15 December 1979).

> Due to the fact that no further uranium development is expected as a result of federal government policies, the town of Jabiru would appear to have stabilised for the time being...As it stands at the moment, I think it has a population of approximately 1200 people (Paul Everingham, Chief Minister, *NT Parliamentary Record* Part 2, 24 August 1983:219).

Plans to turn Jabiru into a large regional centre were rejected by the Ranger Inquiry but the fact that they resurfaced in the Darwin media before construction was complete is a demonstration of their resilience and popular appeal. Only the refusal by the Hawke Labor government to entertain further uranium mining was to be an effective, though interim, brake on the aspirations of those who wished to see diversification and substantial population growth in the town. Many in the NT government, including the Chief Minister, were quite happy in 1978 'just to see the town on the way to being built', given the extensive delays of the early 1970s (*NT Parliamentary Record* Part 1, 28 November 1978:515).

This chapter examines the establishment of the Development Authority to construct and manage Jabiru and investigates the way in which the new community became progressively involved in town governance. It begins with a brief synopsis of events leading to the start of construction activity in mid-1979. Following this, we examine the NT Legislative Assembly Bill which created the Jabiru Town Development Authority in 1979 and some of the important decisions made during the first meetings of the Authority Board. Next, we identify the various steps leading to the introduction of local government in Jabiru and the increasing disenchantment of the NT

Administration with tight commonwealth controls over development as hopes for a start to the new mining prospects faded away. Lastly, is an assessment of competing interests in the early years which we relate to the fourfold classification of intragovernmental tensions introduced in chapter 1. This primarily 'external' view provides the setting for a detailed examination of Jabiru through the views of the residents themselves which follows in Part III of the book.

Events leading to the commencement of construction in mid-1979

Development did not commence immediately as a result of the commonwealth's decision in August 1977 to allow uranium mining in the Region. It was first necessary for local Aboriginal land claimants to enter into consultations with the commonwealth, NLC and mining interests under the NT Land Rights legislation (Altman 1983). Negotiations between the NLC (on behalf of the traditional owners) and the Ranger Partners became something of a *cause célèbre* in the recent history of the land rights movement. Two levels of major disagreement soon emerged: between the mining industry and supporters of the Land Rights legislation (Australian Mining Industry Council, 1978; *NT News* 3 March 1978); and among Aboriginal groups, some of whom suspected a sell-out by senior NLC management (Parsons, 1978; Roberts, 1981; Howitt and Douglas, 1983). Mining industry sources complained of extortionate demands and the need for up-front payments before negotiations could proceed and began a national campaign against the NT legislation (*The Australian* 17 February 1978).

On the other side, there was charge and counter-charge that Galarrwuy Yunupingu, Chairman of the NLC, was being too easily influenced by the commonwealth government (*The Australian Financial Review* 31 October 1978) and that developments like Jabiru would permanently alter the environment of western Arnhem Land to the detriment of Aborigines. There seems to have been considerable confusion among the traditional owners as to the precise nature of the mining proposals, with fears being expressed about a town of 15 000, the disposal of rubbish and the bulldozing of trees. This is not surprising as it was also reported, only two days before the Agreement was signed, that the town plan for Jabiru had not yet been produced at the Ranger negotiations (*Canberra Times* 1 November 1978).

When the s.44 Agreement was finally signed in November 1978 (Figure 4.1) the traditional owners, many of whom were dispersed to places as far away as Darwin, Katherine and Croker Island, 'had a land base to reoccupy and resources to set up a regional organisation

Figure 4.1 Chronology of events leading to the construction of Jabiru

17 May 1977	Ranger Uranium Inquiry Second Report published
25 August	Decision by commonwealth to allow uranium mining to proceed
Feb. 1978	Director of ANPWS appoints consultants to update town plan for Jabiru
1 July	NT assumes self-government
July	Minister for Aboriginal Affairs appoints AIAS to study the social impact of uranium mining in ARR
September	Alligator Rivers Stage I Land Claim granted
3 November	s.44 Agreement under NT *Aboriginal Land Rights Act* signed between commonwealth and NLC
18 December	Director of ANPWS unveils Jabiru Town Plan at Darwin Seminar
3 Jan. 1979	Jabiru Town Development Act passed by NT Legislative Assembly
9 January	Ranger Uranium Project—Government Agreement, Management Agreement and Authority to Mine granted by commonwealth
January	Construction of Ranger mine plant begins
8 February	Cameron McNamara-Minenco appointed by JTDA as Project Managers
5 May	JTDA ready to let first construction contracts
16 July	Director of ANPWS issues licences for work on the town to proceed
23-24 August	JTDA Seminar in Darwin to monitor Jabiru Town Plan
September	Revised Town Plan adopted by JTDA
15 October	Jabiru construction camp operational
December	Commonwealth accepts Peko's offer to purchase its interests. Acquisition to take place through Energy Resources of Australia Ltd (ERA)
Feb. 1980	Peko and EZ establish ERA Pty Ltd to acquire the whole of the Ranger Project
July	ERA becomes a public company

to assist repopulation' (Altman, 1983: 120). They formed the Gagudju Association in 1979 and soon established outstations in the Kakudu National Park with assistance from the NLC. The Aboriginal population in the Region expanded from 60–70 residents in 1975 (at Mudginberri), to 139 in 1980, and 272 by November 1982 (Altman, 1983).

In the meantime the Director of the Parks Service, who now had the responsibility for planning the new town, appointed the consultants A. A. Heath and Partners to update their previous design study. They were required to take into account the major changes arising out of the Ranger Inquiry and prepare a town plan. The layout (Map 10) was based on a list of 25 planning and 30 environmental criteria (Simpson, 1980:93–4). Among the former were important decisions to adopt single family detached dwellings as the main type of housing in the proportions: 609 to 755 houses, 118 to 148 town houses and 110 single quarters. Allotment size was set at 825 square metres which was rather larger than normal in order to 'maintain privacy of sight and sound'. They were to be positioned with an east-west axis to minimise heat

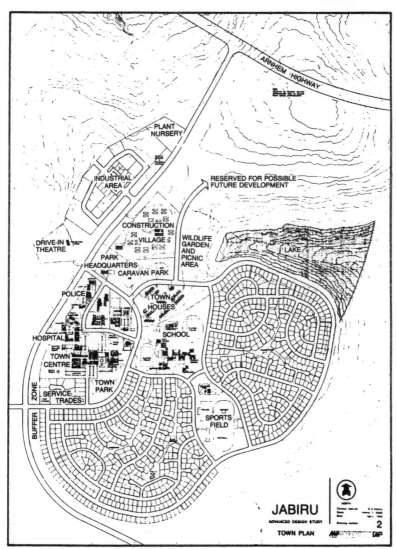

Map 10 *Jabiru Town Plan, 1978.*

load and located with convenient access to walkways and bicycle pathways. The roads were curved to allow houses to be off-set to relieve a regimented effect and the town centre was located at a high point on the site in order to have good visual relationship with Mount Brockman and the escarpment.

The idea for a large artificial lake in the creek bed at the centre of the town layout has an interesting history. Two lakes were originally proposed in the 1972 plans (Cameron McNamara and Partners et al., 1972b), one in the position of the present lake to the east of the town centre and another to the west of the access road from the Arnhem Highway. They were thought of as recreational features because of the long dry season and the lack of any permanent water bodies. The cost of the eastern lake was estimated in 1972 at $250 000, a figure which compares with over $1m when it was actually built eight years later (Jabiru Town Development Authority, 1981a).

The 1973 design study (A. A. Heath and Partners et al., 1973) retained only the eastern lake after adopting a new layout plan and, by the end of 1979, the final impetus which led to its approval was a health report suggesting that varieties of mosquitoes responsible for malaria and encephalitis were prevalent in the area (Commonwealth Department of Health, 1978). The part of the site giving rise to most concern was the natural drainage line leading from the town centre eastwards towards Baralil Creek. Because of this the Health Department required either the removal of existing vegetation and formalising the drainage to eliminate surface run-off, or the creation of a permanent freshwater lake. It would have to be topped-up in the dry season to cover muddy margins and other places where mosquitoes breed (Simpson, 1980:106–7). The former strategy meant extensive concrete drains across the town site and the fact that the more expensive alternative of a lake was chosen says a great deal for the high environmental standards employed by the Development Authority. It is also significant that the choice of a recreational design feature fortuitously became the means of alleviating a serious long-term health hazard.

The 1978 town plan (Map 10) was presented to the major parties involved with Jabiru at a seminar convened by the Director of the Parks Service in Darwin in December 1978. Among those present were representatives of ANPWS, NLC, commonwealth and Territory government departments and the three mining companies. Changes were suggested which included an additional area for future development to the southeast, a change in the residential street pattern with an extension of the pedestrian spines, and relocation of the golf course to the west of the access road (Simpson, 1980:96).

At this point the National Parks Service withdrew from direct involvement in planning the town, considering that it had fulfilled its primary functions of producing an acceptable town plan and the necessary environmental guidelines. Responsibility for building Jabiru was now passed over to the Northern Territory government. In the first months of self-government the new Territory administration prepared the legislation to create a statutory authority to handle the task.

The Jabiru Town Development Authority

> It was agreed that the town be financed, developed and operated by a corporate body comprising the Northern Territory government and the mining companies. After some consideration we decided that the corporate body should be a statutory authority established for the purpose under Northern Territory legislation (Paul Everingham, Chief Minister *NT Parliamentary Record* Part 1, 22 November 1978:365).

The new government's choice was for a development authority although Martyn Finger, Director General in the new Northern Territory public service and an influential voice in the former Department of the Northern Territory, had opposed the idea of a statutory body in his evidence to the Ranger Inquiry (Ranger Inquiry Evidence, 9097–9101). The other possibilities which might have been employed to build Jabiru are not mentioned in the debate on the Jabiru Town Development Bill in November 1978, but it is unlikely that anything else was feasible given the problem of sharing finances among several companies and the need to have a suitable body to acquire the headlease from the Director of the Parks Service. The Authority's managerial functions were considered to be temporary from the outset, to be replaced in due course by local government. We will return to this aspect later in the chapter and it is sufficient to note here that the objective of local government control was also contained in the original plans for the 'open' regional centre in the period before the Ranger Inquiry commenced. The fact that the 'post-Ranger Inquiry' town was to be a closed settlement built on commonwealth land with the complications this must mean for local government received no mention at all in the Inquiry Reports. This was a serious omission when it is realised that many other factors were considered in great detail and lends weight to criticism that important questions about the implementation of 'white' settlement in the Uranium Province were overlooked.

The membership structure of the Authority was to comprise a chairman and six other members, three from the Territory public service and three from participating companies. The first of the latter was the Ranger Joint Venture, though there was provision for later admission of others when they met the criteria for having a significant interest in the town. The Legislative Assembly debate on the Bill reflects considerable bi-partisan support though Bob Collins, Leader of the Opposition, emphasised that the Jabiru Town Plan was inexplicably deficient in one important respect. No mention was made in the body of the report about the future needs of Aborigines in the town:

> ...I again searched in vain for any costing in regard to Aboriginal living areas. There is none. I looked under housing and accommodation; it does not receive a mention there. I looked through the entire costing and I could not find a mention of it anywhere...Perhaps after they build their $450 000 church and fill it up with christians, they can take up a collection for the Aboriginals (*NT Parliamentary Record* Part 1, 28 November 1978:503).

This was a matter soon to be addressed by the Development Authority in their review of the Town Plan provided by the Parks Service in early 1979. There was also considerable comment about the necessity to control and restrain the new residents to ensure their lifestyle and behaviour did not conflict with the location in the National Park. The banning of cats and various exotic plants which this entailed appeared to politicians on both sides of the House to be a justifiable and necessary fact of life. Only one speech drew any close comparisons between the plans for Jabiru and other new mining towns in the Territory. Mrs Lawrie, Member for Nightcliff, compared the lifestyle and expectations of people in Nhulunbuy and Alyangula (Groote Eylandt), observing the latter to be immensely preferable because of the social differentiation and poor design of company housing in Nhulunbuy. At Alyangula, by contrast, the emphasis was on natural ventilation and cooling and the preservation of local flora. It should be noted, however, that this situation was soon to change when unions succeeded in obtaining full airconditioning for the GEMCO houses on Groote even though they were not designed for it.

Great care was taken in Jabiru to design housing suitable for airconditioning and to minimise segregation and enclaves but, as we shall see, achieving similarity in external appearance of dwellings does not prevent considerable differences in entitlements of other kinds. The relatively disadvantaged status of government employees is endemic in mining towns and cannot be removed by physical design features alone when important recurrent items such as water and electricity charges are ignored.

Construction begins

The Inaugural Meeting of the Development Authority was held in Darwin on 16 January 1979. E. J. Simpson, from the Special Development Projects Section of the Chief Minister's Office, was the first Chairman and other public servants present came from the Department of Transport and Works (C. C. Russell), the Treasury (W. J. F. Hull) and the Department of Community Development (J. R. Larcombe). Three company representatives from the Ranger Joint Venture were D. T. Woods (General Manager of Ranger Uranium

Mines), A. D. Blain (EZ Co. of Australasia) and W. S. Arthur (alternate for A. H. McIntosh, Peko-Wallsend). The early business demonstrated immediately the advantages of a joint public and private sector partnership, with funds coming from both sides to enable work to proceed. Ranger made a $2m advance to cover establishment and operational costs and a $100 000 loan was granted by the Northern Territory Treasury for initial working capital. Company members also undertook to prepare a draft system of financial control which could be put to immediate effect.

A. A. Heath and Partners were appointed to finalise their earlier town plan incorporating changes suggested at the December 1978 Seminar in Darwin (Map 11), but it soon became clear that areas for future residential growth were too far removed from the town centre and other amenities and presented servicing problems. Accordingly, another planning workshop was held in Darwin in August to suggest further solutions and resulted in a layout design which was used to construct the town (Map 12). It incorporated one major change: the relocation of future residential development to the west of the main access road, making the town centre the nucleus of the whole development. Minor amendments included the relocation of the high school, sports club and swimming pool. The area reserved for Aboriginal use was changed to a location north of the lake from its previous position close to the police station—an aspect of the plan which had been criticised by the Leader of the Opposition in the Legislative Assembly debate on the Development Authority some months before.

Delays were experienced in beginning construction work because of difficulties in obtaining approval under the provisions of the National Parks Act. Not surprisingly, the Act did not provide for town construction in a national park and was an early example of many legislative and procedural changes which had to be made to bring the Ranger Inquiry 'solution' into effect. Cameron McNamara-Minenco Joint Venture were appointed Project Managers at the Second Meeting of the Authority in February in a decision which reflected their prior involvement in the early plans for the town and the fact they already possessed an office in Darwin and a Territory identity. The Authority was ready to call for headwork tenders by April but could not proceed until June because of delays in the amending legislation to the National Parks Act.

Financing the town

The newly established harmony between government and company members of the Authority was soon threatened by disagreement over the financing of the town. The crux of the problem was whether headworks and other infrastructure should be paid for by the government,

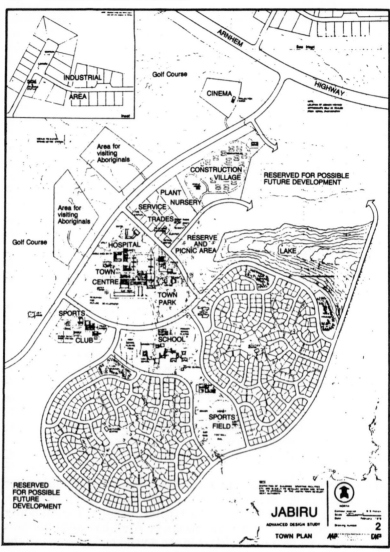

Map 11 *Jabiru Amended Town Plan, 1979.*

recouping such costs when sub-leases were issued to the companies, or whether the private sector should pay for their share of all expenditure from the outset. Added to this was considerable confusion about cost-sharing arrangements between the three companies, given the reality that only one of them had secured an authority to mine.

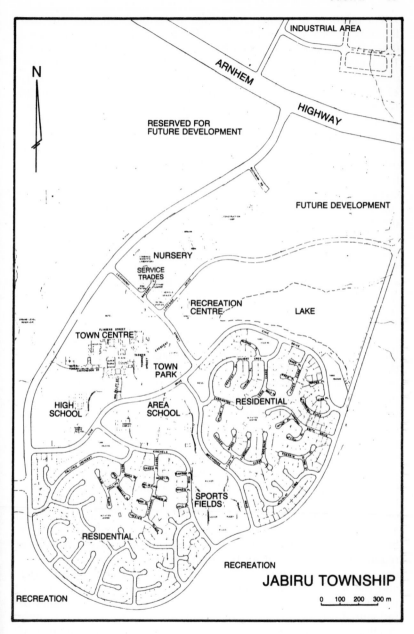

Map 12 *Jabiru Final Town Plan, 1979.*

The origin of the disagreement was in the Chief Minister's Second Reading Speech on the Jabiru Town Development Bill where he stated that, 'It is expected that the initial source of money will be loan funds. The costs of the Authority will be recouped by premiums on subleases of land and by rates and charges' (*NT Parliamentary Record* Part 1, 22 November 1978:366). On the strength of this, two Members of the Authority went to Canberra in February 1979 to arrange the necessary borrowings from the commonwealth only to be told that Ranger should be prepared to provide full funding for the town from the start of construction on 1 July 1979. Company reaction was swift: as far as they were concerned the Authority should finance the town plan from funds already received and borrow the rest to pay for headworks and infrastructure, subject to periodic cash contributions from the three companies as they took up their housing entitlements; the Authority should also finance and own community facilities until transferred at some future date to a municipal council; the companies would enter into agreements with the Authority for provision of town facilities according to a formula based on *pro rata* shares of residential sites; and each participating body was to be responsible for financing, construction and maintenance of their own employee housing (Telex from General Manager of Ranger Uranium Mines to Chairman of JTDA, 28 March 1979).

Neither the commonwealth nor Territory governments accepted these proposals and determined that Ranger would pay for most of the work in advance by instalments, with federal guarantees only for those facilities in excess of Ranger's own requirements. The Ranger members were unhappy with this decision, feeling that they would end up by paying a great deal more than they should, but had no option but to agree to the terms laid down. To have decided otherwise meant even greater commercial costs in delaying the entire project.

Cost sharing remained the chief area of dispute for some time with Ranger, in particular, being anxious to clarify the split between themselves and government and the determination of an equitable formula for spreading charges among the three companies. Final resolution was not to be reached until 1985 (see our conclusions) but a provisional agreement was made that Ranger would pay for the headworks and most town facilities except those in excess of its own requirements but needed for the other companies. Commercial buildings in the town centre would be a joint responsibility of Ranger and the Authority on behalf of the others. A formula was needed which would indicate each share according to company occupancy of residential lots. This meant converting the housing required (family dwellings, townhousing, single quarters) to an equivalent number of 'standard' residential blocks for purposes of comparison. Thus two townhouses or single quarters could equal one standard residential block, and so on, with calculations based on the latest population forecasts by each company. Provisional agreement was reached in these terms together with an overall split between Ranger and the Authority on the understanding that further talks would iron out the details in the following year. As we have already indicated, the failure by Pancontinental and Denison (successor to Noranda) to gain an authority to mine delayed resolution long beyond this date.

The staging of residential development

Another aspect of town construction which was subject to consider-able uncertainty was the staging of residential development and the spatial distribution of housing lots among companies and government agencies. Mining town literature suggests that various negative social consequences flow from the creation of employment-based enclaves but are a possibility which must be balanced against cost penalties arising from dispersed and mixed development (Lea, 1984). It is clear, for example, that reserving vacant and serviced lots for the two other companies would add considerably to the infrastructure cost of the town in the short term. Ranger members of the Authority called for

Table 4.1 Cost estimates for three different ways of accommodating Ranger personnel in Jabiru

	$'000s		
	Random distribution	Concentrated distribution	Cluster distribution
Roads & minor drainage	7 359	4 084	5 624
Drainage	4 050	1 650	3 950
Sewerage	3 321	2 822	3 122
Water supply	4 322	3 722	3 982
Fencing	150	150	150
Recreation	2 325	2 325	2 325
Elec. & cables	3 059	1 980	2 495
Lake	1 307	1 307	1 307
Housing	41 380	41 380	41 380
Town centre	9 438	9 438	9 438
Other components	5 240	5 115	5 165
Total	81 951	73 973	78 938

Source: Jabiru Town Development Authority (1979)

a costing exercise showing the variable costs of housing their own personnel according to three possible distributions (Table 4.1). The random and clustered possibilities include the opportunity to mix employees of the three companies at a later date, whereas the concentrated distribution does not.

The economics of sewerage and drainage systems decided the Authority in favour of initial development of 286 blocks located at the northern ends of the two residential nodes: 99 were allocated to Ranger and scattered over the two main areas to be developed in a random manner; 87 were similarly scattered and allocated to other bodies such as government, private sector and the Authority itself; the remaining 100 blocks were considered to be 'spare' and would be left as such until April 1981 before a decision would be made to allocate them to the other mining companies, or to use them for the needs of existing bodies in the town. As matters turned out some of these spare blocks were allocated for use before 1981 and additional ones were created making the number up to 140 (Map 13). At the time of writing (early 1985) they are still undeveloped and have attracted considerable interest charges on top of the capital originally borrowed from the commonwealth. It represents a sum of several million dollars forming the bulk of the Authority's outstanding debt.

Uncertainty also extended to decisions about the capacity of the headworks and forced the Authority to make a judgement on the eventual size of the town on economic and engineering grounds. The assumption that a 'Ranger only' town of about 2250 was likely did not alter the necessity to over-design certain facilities which could only be enlarged at great cost later. Water mains and sewerage plant were made large enough to serve a town of 6000 and the water treatment plant and storage tower for 3000. The overall approach was to design as many components as possible to the lower figure and make provision for augmentation, if this could be achieved without cost penalties at a later date. The borefield for permanent water supplies was located at Nanambu Creek, some 20 km west along the Arnhem Highway. Initial supplies came from bores along the Baralil Creek but these were unsuitable for long terms needs. Elevated water storage was necessary to ensure reasonable service pressures in a relatively flat topography, four designs being considered for a water tower costing about half-a-million dollars, with a decision being made in favour of reinforced concrete with a conical shape.

Housing

Some of the most important early decisions concerned the design, distribution and costs of housing. The companies were initially inclined

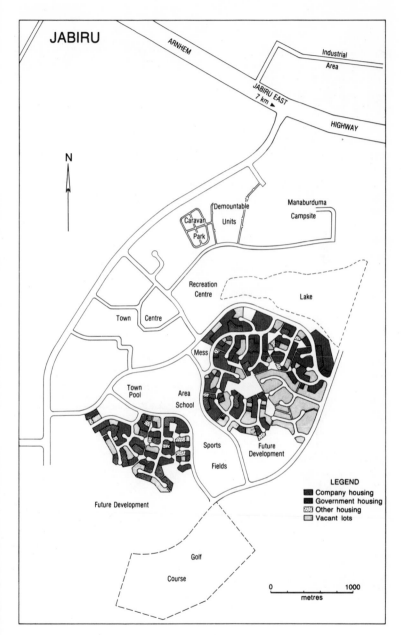

Map 13 *Distribution of housing types in Jabiru, 1984.*

Housing types *High set house, Jabiru East* (above left); *Ranger house, Jabiru* (below left); *undeveloped subdivision (vacant lots), Jabiru* (above).

to handle their needs independently, subject only to basic planning and standards control by the Authority but Ranger ultimately adopted the designs and recommendations of a housing report by the Project Managers (Cameron McNamara-Minenco Joint Venture, 1979a). An architect and engineer were sent to Weipa, Nhulunbuy (Gove), Batchelor and Darwin to examine designs and construction materials, coming up with a range of high and low set housing alternatives. The company's decision on technical and economic grounds was for low set (single-storey) housing with verandahs on all sides and a large covered area for car parking and other purposes. A comparison of the advantages and disadvantages of single- or double-storeys determined in favour of low set designs if full airconditioning was included, and the fact they were five per cent cheaper led to their acceptance. This was not to be a popular decision for some residents as the results of the community survey (chapter 5) indicate.

Another important decision was for uniformity in housing construction materials regardless of ownership. This was a departure from the practice in other mining towns in Australia and decided upon in order to minimise obvious distinctions between company and non-company residents. The objective was greatly assisted by the generally similar styles of housing chosen for both groups. Ranger dwellings came in four 3-bedroomed and two 4-bedroomed designs and were superior to the other accommodation in several respects being fully, rather than partially, airconditioned and possessing higher quality internal features. Double-storey townhouses for couples without children and single persons quarters grouped in 22 blocks of 6-room units, were also provided for company personnel.

Planning for singles in Jabiru was based on an alternative to the accommodation found in earlier mining towns which tended to segregate single employees in their own enclave in town or at a site outside closer to the mining operation. The Project Managers suggested blocks of six-room motel-style units for both company and government employees scattered around the northeast residential node with the proviso in the case of the company accommodation that they should be in walking distance of the mess. No more than 18 singles were to be grouped on any one site to minimise any adverse effects of large numbers. Most of the single person blocks were located at the head of culs-de-sac in order to maximise opportunities to use the pedestrian walkway system. Questions about this scheme and how it operates were included in the community survey and are considered in more detail in chapter 5. Single accommodation for government employees differed in being self-contained and was also scattered throughout the residential area, but was not restricted in location to the walkway system.

The first 100 Ranger family houses cost $71 490 each, excluding land and site development costs, based on contract price (JTDA, 1981b). Figures for Ranger townhouses and single persons quarters were $77 111 and $21 272 respectively. The first contracts for government family housing and single quarters resulted in comparable costs of $69 400 and $40 500 each (JTDA, 1981b). The much higher figure for the latter was because the government single quarters are larger and fully self-contained. These figures reflect the reality that housing construction in the Australian tropics is generally double the cost found in the southern states (Commonwealth of Australia, 1974) and is due to high labour charges, materials, freight, plant hire, communication problems, and so on (Newton, 1977). Fixed on-costs per house, for example, were estimated to be of the order of $3850 for Jabiru, being made up of cartage from Darwin, over-award payments and camp charges (Cameron McNamara-Minenco Joint Venture, 1979a).

Street names

The first idea for street names in Jabiru came at a Development

Authority meeting in March 1979 when it was decided to call them after birds following the practice already begun in Jabiru East but this was subsequently changed the following year. The first batch of names were chosen from the members of early exploration parties in northern Australia, in particular those of Leichhardt and Stuart: Calvert, Brown, Phillips, Roper; and Kekwick, Thring, Auld, King, Waterhouse, Frew, Nash and McGorrerey. In addition, names of pastoralists were suggested such as Paddy Cahill and Lewis who lived in the Region at the turn of the century.

A second batch was chosen from three additional sources as well as certain obvious general features such as Lakeside and Civic Drives. First were some Gunwingga words chosen for their brevity and agreed to as being suitable by the local Aboriginal community: Kambuk (freshwater crab), Kinga (saltwater crocodile) and Miri (oar). Second were some well known topographical features such as Mount Brockman and the Magela Creek; and third were personalities and items of historic interest taken from Cole's (1975) book on Oenpelli: Dyer (after Rev. A. J. Dyer who, with his wife, was a pioneer at the Mission between 1925 and 1934), Pera (the Dutch vessel which accompanied the *Arnhem* in the exploration of the Gulf of Carpentaria in 1623), Quilp (the Aboriginal stockman whose life was saved by Paddy Cahill) and Morey (the Territorian policeman who carried out extensive patrol work in the Top End from 1927 to 1947).

Aborigines in Jabiru

It was clear by mid-1980 that plans were needed for the area of town set aside for Aboriginal occupation and zoned for this purpose in the National Park *Plan of Management* (ANPWS, 1980). This showed two areas amounting to 15 ha in the northern part of the site between the Arnhem Highway and the lake. They were originally areas designated for Aboriginal visitors rather than permanent occupation and were required to be shielded by a 'buffer-zone' of vegetation. It was uncertain, however, whether Aboriginal mine employees might also wish to live here rather than in conventional sub-divisions and lengthy negotiations with the NLC and Gagudju Association led to an application for commonwealth funds to develop the sites. The initial agreement was for six simple shelters and two serviced ablution blocks, with the $144 000 cost of the buildings and service connections and $74 000 for road and drainage to be borne by the commonwealth (Jabiru Town Development Authority, 1983). Provision was also made for Aboriginal use of 12 additional allotments of a conventional kind in the main part of town when the need arose.

A small Aboriginal group had shifted from Mudginberri to the mine

Aborigines in Jabiru *Restricted entry sign to Manaburduma Camp, Jabiru* (above left)*; Aboriginal shelter, Manaburduma Camp, Jabiru* (below left)*; sign, Kakadu National Park* (above)*.*

camp at Jabiru East when facilities became available there in 1979. Concern was expressed in Authority meetings that the same people would move to the new town as services such as the hospital were built, without suitable housing being ready for them (see evidence to the House of Representatives Inquiry into Fringe-Dwelling Aboriginal Communities, 1982:1757–86, for further discussion). At first it was thought that they would form a temporary camp but, by 1983, it became clear that a permanent community was living in the six shelters. It was thus necessary to upgrade the housing provided earlier, and individual washing facilities, lighting and power were provided to the houses and ablution blocks at a cost to the Commonwealth of a further $140 000.

A problem for the Authority was whether to control this settlement through issuing a sub-lease to the Gagudju Association or to request the Northern Territory Housing Commission to look after the dwellings. Ultimately, neither of these possibilities was chosen and the Authority decided to retain control over the camp because the Gagudju Association was unwilling to agree to certain limitations in the proposed sub-lease which were bound up with obligations under the head lease with the Parks Service. The camp has not been dedicated to the public in order to avoid problems under the Northern Territory *Liquor Act* and no charges have been made to occupants for the use of the services provided pending the introduction of By-laws (Jabiru Town Development Authority, 1985). This has resulted in an annual charge to the ratepayers of Jabiru (largely ERA Ltd) of some $100 000 which has not been recouped from the commonwealth. The latter was met on a once-only basis by the Northern Territory government in 1983/84 but remains as a future problem for the new Town Council. The proposed By-laws will impose a camping fee and contribution towards services and are designed to protect the campers from unwarranted interference (Jabiru Town Development Authority, 1985). Numbers in the camp (later called Manaburduma) have not increased much above 50 persons in the four years since the town was occupied and there is no evidence of unauthorised camping elsewhere in Jabiru.

Several other issues such as the design of the town centre and the vexed question of housing the private service sector also occupied much of the Authority's time in its first two years of operation, though there is insufficient space to address them here. The Development Authority was successful in managing the construction and occupation of a $100 m town in three years and was the first body of its kind in Australia to have built a mining settlement in this way. It was also the first town to have been largely financed by one company on the basis of retrospective cost sharing and the complexities overcome in doing this may well become the norm in multi-company involvement in remote resource developments in the future.

Of central importance to our discussion, however, is the way in which the new residents became involved in town affairs after the first occupants arrived in July 1980. It is an interesting case study of the birth of a new rural municipality set against a backcloth of government and company tensions in the years immediately after self-government in the Northern Territory.

Town government

We have noted previously that the Ranger Inquiry limitations on developing Jabiru were to change the basis upon which the town might hope to attain municipal status. There was no freehold property ownership and housing in the community was to be allocated only to households directly or indirectly associated with the mining industry. Added to this, it was clear by now that the town would be largely privately financed, hardly an optimistic scenario for widespread popular participation in municipal affairs. Yet, surprisingly, the notion per-

Jabiru town centre under construction, 11 September 1981 (above) and 30 April 1982 (below).

Jabiru town centre *Shops* (above left); *tourists arriving at Jabiru town centre* (below left); *Kakadu tourist congestion at Jabiru town centre* (above).

sisted that the Development Authority was a temporary institution to be replaced by normal local government in the near future. This was confirmed by the Chief Minister in the debate on the Jabiru Town Development Bill in November 1978 and shortly afterwards by the Chairman of the Authority.

This persistent government-led push to bring local government to Jabiru is one of the most interesting features of the case study and represents a major departure from either the 'bottom-up' popular origins of the early Victorian examples we have considered, or the company-initiated 'normalisation' found in the Pilbara. Neither Jabiru residents nor the mining companies actively sought local government in the early stages of the town's development and it seemed that it was only the Northern Territory Cabinet which kept the idea alive. We have already suggested that this is best explained in terms of political goals which were generally external to the town and its immediate priorities and were consistent with the actions of a new Territory administration anxious to create the pre-conditions for introducing local government to as many of the smaller centres as possible.

The Everingham Cabinet was a firm believer in decentralising administrative responsibility of the kind which they had won for themselves on the attainment of self-government. They were faced with a situation in the Gove Peninsula and the Uranium Province where two of the richest development prospects in the Top End remained outside their control in the hands of the commonwealth and Aboriginal groups. This was all the more galling because the Territory had extensive administrative and servicing responsibilities in the two regions and the residents of the towns were part of a rapidly growing new electorate. The ruling Country Liberal Party (CLP) lost the seat of Gove to the Labor Opposition in the 1980 elections and it seems certain that a postmortem identified the authoritarian control exercised by NABALCO, the bauxite company, and the absence of elected local government in Nhulunbuy as contributing factors. Property ownership, as we observed in chapter 2, has been the cornerstone of political activity at the local level in Australian municipal history and it is likely that political parties at the conservative end of the spectrum, such as the CLP, would benefit most from a move towards normal local government and the rights to private ownership that go with it.

Thus the key to understanding the chain of events which led to the introduction of local government in Jabiru in mid-1984 (Figure 4.2) is the political scene in the Territory in the years immediately after self-government and the ambitions of its ruling party. Local factors were also to play their part in the timing of events but did little to alter the long-term strategy which had already been determined in Darwin. The unfolding of this strategy which led ultimately to the 1984 Council elections in Jabiru fell into two phases: a three-year period of increasing government pressure on the Authority to introduce local government; followed by a two-year transition with an elected advisory council. A third and ongoing phase is witnessing the winding down of the Authority and the assumption of administrative functions by the new Jabiru Town Council. Full local government in Jabiru will not be achieved until the Council acquires ownership of the headworks and municipal assets and power to set the rates.

Pressure from Darwin

The Jabiru Town Development Act of 1979 established the primary functions of the Authority to develop, administer, manage and control the town but did not include reference to a future transfer of powers to local government. It was an assumption, as we have noted, which was contained in the Chief Minister's Second Reading speech but was not formally recognised until negotiations took place with the commonwealth in 1979 about the head lease over the town site. The lease contains a provision (s. 17) for the transfer of powers to a future

Figure 4.2 Chronology of events leading to local government at Jabiru

22 Nov. 1978	Jabiru Town Development Bill 1978 debated in the Northern Territory Legislative Assembly. Passed 3 January 1979
13 Dec. 1979	JTDA Chairman states objective to hand over to local government within three years with an advisory council in the interim
25 July 1980	First residents move into Jabiru
3 Feb. 1981	Chief Minister instructs JTDA (s.4(4)(b) *Jabiru Town Development Act*) to form a citizens' advisory group by the end of 1981 and elected town council by mid-1984
4 March	Chief Minister announces that JTDA is moving towards establishing a citizens' advisory group
2 June	ERA writes to Chairman of JTDA giving reasons why full local government should not be introduced for 10 to 15 years
29 June	Jabiru Town Development Amendment Bill defining local government powers and functions of the Authority passed in Legislative Assembly head lease over town site signed with Director of National Parks Service
17 August	Senior public servants meet in Darwin to discuss local government for Jabiru
11 Mar. 1982	Private Member's Bill introduced by Terry Smith (Labor, Milner) to allow several residents to be elected to membership of JTDA. Debate adjourned
31 March	Meeting in Jabiru between JTDA and 35 invited residents to discuss the idea of citizens' advisory committee and date for elections
2 April	Northern Territory Cabinet endorses formation of an advisory council and date for elections on 22 May
15 April	Public Meeting attended by more than 100 residents of Jabiru addressed by Chief Minister on question of citizen participation in town government
22 May	Advisory Council elections held
24 May	K. T. Danielson, E. M. Main, D. N. Atkinson, G. A. Court and D. B. Green declared elected
2 June	Jabiru Town Development Amendment Bill passed allowing creation of an Advisory Council. R. McHenry, G. Stolz and A. H. McIntosh nominated members
26 July	Town of Jabiru officially opened by Hon. Doug Anthony
27 August	Dr Atkinson transferred from Jabiru and vacancy declared on Advisory Council
1 June 1983	Chief Minister requests Advisory Council in consultation with JTDA to advance proposal for local government
24 August	Chief Minister announces in the Legislative Assembly winding down of JTDA and the plans to devolve local government functions to the Advisory Council
19 October	Chief Minister announces in the Legislative Assembly a proposal to amend the *Jabiru Town Development Act* to allow the delegation of municipal powers, other than determination of the rates and employment conditions of municipal staff, to a new Jabiru Town Council
29 Feb. 1984	Jabiru Town Development Amendment Bill read for second time in Legislative Assembly. Passed on 7 June 1984
26 May	Jabiru Town Council elections: E. M. Main, G. A. Court, D. G. Green, M. Gray and N. Rillstone elected
1 July	Jabiru Town Council established

municipal authority and indicates that the commonwealth, in the form of the Parks Service, was fully aware of the possibility at an early stage. Whether it recognised the speed with which events were to develop is much less likely.

The first step in the process was the appointment of a Town Manager who would pass on the views of a residents' advisory committee to the Authority. This requirement was included in the job description for the Manager's position when it was first advertised early in 1980. However, no action was taken to establish such a citizens' group until the Authority received a letter from the Chief Minister in February 1981 ordering it to do so by the end of the year. In addition, they were instructed to establish a form of fully elected local government by mid-1984.

There was no obvious pressure for this from residents, the first of whom had moved in eight months previously, or from other sources for this initiative and the most plausible explanation is the political events in Nhulunbuy at the end of 1980. It is likely on the basis of that experience that the motivation had more to do with preventing a similar authoritarian company presence in Jabiru than it had with counteracting commonwealth influence in the Uranium Province (NT Cabinet Decision 1806, 1981). Energy Resources of Australia (ERA), the new company which had been formed in 1980 to acquire the Ranger prospects, responded to the directive with a long list of reasons why local government should not be introduced for at least 10 to 15 years. They expressed fears that a local council composed largely of temporary residents might be able to unduly influence the spending of funds it did not raise. There was also concern that borrowings to construct the town were subject to an eight-year floating charge over their assets and that it would not, therefore, be possible to transfer ownership of headworks and facilities until the loans were repaid in full. Such assets were seen as shareholders' funds and part of the essential infrastructure of the mine itself, a point emphasised in the political economy model of mine development.

Discussions between the Authority and senior Darwin public servants involved in the town took place during 1981 with a view to reconciling the differences. It soon became clear that the problem of the floating charge could be removed if the Authority was empowered to hold company assets in trust until such time that the loans were repaid. In addition, it would be possible for Northern Territory government agencies to assume responsibility for maintaining the headworks and power supply on the transfer to local government and thus ensure these vital parts of the infrastructure were looked after until the expiry of the mining lease.

Amendments to the Authority's Act were passed in the Legislative

Assembly in June 1981 enabling it to perform new functions such as the creation of by-laws and a rating system. This was also seen as a move facilitating the future transfer of powers to a municipal body by having the necessary local government machinery already in place. The Opposition welcomed the changes but expressed some doubts whether local government in Jabiru would actually be allowed any freedom of action given the Territory government's record in interfering with other councils (*NT Parliamentary Record* Part 1, 10 June 1981:1127).

Despite these manoeuvres in Darwin, progress towards active citizen participation itself was not very noticeable in Jabiru and no advisory council was in place at the end of 1981. The reasons for this were probably due to continuing construction activity and the fact that signs of community discontent with the managerial style of the Authority did not reach the ears of Darwin politicians until early in 1982 (Lea and Zehner, 1985).

The Jabiru Town Advisory Council

Evidence of community dissatisfaction with the Authority began to surface in the editorial columns of the *Jabiru Rag* in February and March with complaints about the probable downgrading of the hospital to a clinic and the general lack of information about major development issues (*Jabiru Rag* 22 February 1982). Organisations with interests in the town such as the Gagudju Association complained of a breakdown in communication between the Development Authority and the Town Manager: 'We go to the town manager and work something out and that's when we find out he says you have to go to the Jabiru Town Development Authority...there is a breakdown of communication of decision making' (personal communication: Ross Hebblewhite, General Manager Gagudju Association, 1983). Individual residents cited similar problems: 'It was very hard to talk to the man in charge...was a one-man show' and 'I had difficulties previously with the Town Manager...' (Jabiru Community Survey, 1984). The Northern Territory Labor Party happened to be investigating the establishment of a branch in the town at this time and decided to stir things up for the CLP by putting an Amendment Bill to the March Sittings of the Legislative Assembly with a proposal to include three elected residents on the Authority Board (personal communication: Terry Smith, Member for Milner). This action 'put the wood under the pot' according to Paul Everingham, Chief Minister at the time (personal communication), and led directly to two meetings in Jabiru between the Authority, politicians and residents about the town's problems and future administration.

The first meeting was closed to all but 50 invited residents who were, in the opinion of the Authority and ERA, 'public spirited and representative of local opinion'. Some 35 attended the gathering held in the Jabiru Child Minding Centre on 31 March 1982 and it soon became obvious that the failure to hold an open meeting was a tactical mistake. The Authority had hoped to receive endorsement of a proposal to hold elections on 22 May for an advisory council of four elected members and three appointees (one each from ERA and the government with the Town Manager in the Chair), which was the same composition as the Nhulunbuy Town Board. Many present questioned why the attendance was restricted and called for a full public meeting before any decisions were made. They also resented the 'cut and dried' nature of the Authority's proposals and demanded that the number of elected members of the council be raised. The editors of the *Jabiru Rag* were so disgusted with the way the affair was handled that they resigned their positions on the newspaper, saying 'we find we cannot make any comment at all about the meeting that was held without offence...' (*Jabiru Rag* 5 April 1982). Interestingly, one of them was Eric Main, a clergyman, who was later to be elected as the first Chairman of both the Advisory and full Town Councils.

The public meeting was duly arranged two weeks later and was addressed by the Chief Minister who had realised that his own presence was needed to restore credibility in the Authority. Although the meeting was to be the public forum for a decision, the Northern Territory Cabinet had already decided in secret, on 2 April, to go ahead with the elections on 22 May. By all accounts the Chief Minister dominated the proceedings and the meeting ratified his proposal for five elected members from within a boundary defined as a 10 km radius of the Jabiru police station. Three members were to be nominated by himself with the Chairman to be chosen from the elected representatives (changes in mid-1985 led to a reduction in the number of nominated members to one and an increase in elected representation to seven). Out of the five elected from a list of 15 candidates on 22 May, only one was a mine company employee and another, who owned a small business, was the spouse of a member of the mine management. Although three-quarters of households in the town lived in mine housing, there are at least three reasons why a majority of non-company members were elected. First, the mine households could have been secure in the knowledge that their interests would be looked after by the Company members of the Development Authority; second, that an ERA employee was one of the appointed members of the Advisory Council in any case; and, third, that it was the non-company people in the community who were most likely to benefit from gaining a voice in town affairs (Lea and Zehner, 1985). Direct evidence to support the latter point of view is contained in the editorial of the *Jabiru Rag* for 8 March 1982, which lists six items judged to be to the comparative disadvantage of public servants in Jabiru. They include factors such as inferior housing, high rents, partial airconditioning and, for all residents, the inability to purchase housing in the town.

The Territory government played down these signs of discontent in the town, though they did agree that the Authority might have been 'lacking in sensitivity' and that, in any case, it was really a construction institution rather than management authority. Other comments made in the Legislative Assembly at the time were not very sympathetic to the issues being raised and echoed a familiar response to the circumstances found in some new company mining towns:

> . . .it could be said that one of the problems of Jabiru is that there is not a tremendous amount for people to do. Everything has been done for them. Their gardens are virtually set up ready for them and so on. I guess that time, especially in the wet season, has preyed on the minds of some people rather unduly (Paul Everingham *NT Parliamentary Record* Part 1, 27 May 1982:2333).

There was, of course, some truth in these assertions. One resident responded during the 1984 Community Survey that there is: 'Too much leisure. . .leads to drink. . .leads to social problems. Everything is done for everybody. It's all provided. I'm a member of a service club. . .it's damned hard to find any worthwhile projects to do.' But it was the weather, that familiar cause of tropical ennui, in concert with an overgenerous Development Authority, which were held chiefly responsible for yet another item in the long list of difficulties confronting white settlement in the North.

The first meetings of the Advisory Council were concerned with various complaints raised by the elected members and directed towards their appointed member colleagues (the Chairman of JTDA, the Northern Territory Co-ordinator General and a representative from ERA) and the Town Manager who attended as a non-voting Secretary. There was confirmation of the perceived inadequacy in communication between the Authority management and residents as well as specific complaints about matters such as medical care, the community hall and the use of pesticides in mosquito fogging. The town doctor, the Councillor most vocal in his criticism of medical resources, was shortly transferred from the town by the Northern Territory Department of Health, leading to questions being raised in Darwin about the circumstances of his removal. The Chief Minister denied in the Legislative Assembly that it had happened for political reasons and stated that the transfer was part of the normal rotation of medical staff in the Northern Territory public service (*NT Parliamen-*

tary Record Part 2, 17 August 1982:909). But suspicions remained in the town that the doctor had been removed because his comments were politically damaging.

In many respects the first year of Advisory Council activity were a marked success with numerous community matters receiving attention and activities organised which could not have been adequately handled by the Authority. Chief among these were the organisation of a Community Awareness Week and events to celebrate the first anniversary of Jabiru in June 1983. Responses to the Community Survey revealed that some residents were not so convinced and that many were not aware of the Council's existence or role. Typical of these answers to a question about what JTAC did were: 'an interface between the community and JTDA... a toothless dog that added more red tape to the system'; and 'very little. Did a lot of talking. Didn't achieve very much. A con to make people think they were running their own affairs.'

Unfortunately, the uncertainties which accompanied the election of the Hawke government in Canberra in early 1983 had an immediate effect on ERA. As a uranium mining company it had good grounds to fear the consequences of a Labor administration and became noticeably less generous in accepting proposals to finance further development of community facilities. Confirmation in 1984 that the mines at Jabiluka and Koongarra would not be allowed to proceed did not really come as a surprise to the town but, because the expansion of Jabiru to accommodate additional companies would not now be necessary, it pronounced a death sentence on active involvement by the Authority in further community developments. Government and ERA were now agreed that it should be wound down and most of its functions transferred to a Town Council. The latter was to come into being after the biennial Advisory Council elections in mid-1984 (*NT Parliamentary Record* Part 2, 24 August 1983:220).

There were no illusions about the complexity of introducing local government in Jabiru where the most secure form of property ownership could only be on the basis of a sub-lease shorter than the 40-year head lease with the National Parks Service. Even the Chief Minister got the details wrong in a Legislative Assembly speech at this time, assuming that the town site was on Aboriginal land leased by the Parks Service to the Development Authority, making any future residential lease in the town the sub-lease of two others with better title (*NT Parliamentary Record* Part 2, 19 October 1983:366)! Although matters were slightly less complicated than this due to commonwealth ownership of the town site, there would nevertheless be no personal property ownership in the normal freehold sense and the mining company would still be responsible for most rates and charges. Table 4.2 shows

Table 4.2 Ratepayers in Jabiru 1984

Ratepayers	$	%
ERA	384 508	62.8
NT government		
NT Housing Commission	121 628	19.9
NT Police Force	1 703	0.3
Commonwealth		
OSS	56 554	9.2
Dept of Admin. Services	9 552	1.5
ANPWS	6 455	1.1
Telecom	6 169	1.0
Dept of Health	1 683	0.3
Bridge Autos, Jabiru	12 523	2.0
Westpac Banking Corp.	11 355	1.9
Total	612 130	100.0

Source: Jabiru Town Development Authority (1985:36)

that there were only 10 ratepayers in the town in 1984 with two-thirds of the total income of $612 000 coming from the mining company and most of the rest from Northern Territory government or commonwealth sources. This narrow rating base will mean that Council income will be insufficient to pay-off existing Development Authority debts were they to be transferred at some time in the future (see our conclusions for further discussion).

Legislation to give the new Town Council the local government powers of the Authority was introduced in the Legislative Assembly in February 1984. Two important powers were to be retained by the Authority: responsibility for setting the rate and control over the establishment and conditions of employment of council staff. The Authority would remain for the time being to carry out these tasks and maintain the ownership of headworks and other town facilities in trust for the mining company. The Bill also contained provisions to do away with the three appointed members of Council at a future date, so could be said to institute a form of transitional local government tailored to the peculiar circumstances of Jabiru. An interesting feature of the subsequent debate was the reference drawn to the new township of Yulara near Ayers Rock and the advice of the Minister for Conservation that it should look to Jabiru as a possible model for its own use (*NT Parliamentary Record* Part 1, 7 June 1984:514).

The bitterness felt by Territory politicians over the unwillingness of the commonwealth to give up its considerable powers over development at Nhulunbuy and Jabiru was never far below the surface, as illustrated in this example from the same debate:

> We have to deal with the greatest tangle of federal bureaucratic mumbo-jumbo, gobbledegook and red tape in any part of this country—a country that is held back by a city that is dedicated to red tape and bureaucracy, a city that is peopled largely by citizens who have nothing better to do in many cases than to stop things or slow them down or get their fingers into pies that do not concern them when the rest of the country is out and trying to make a buck to pay the tax dollars that keep the city of monuments in existence...that is the situation with local government in the Northern Territory. Local government in the Northern Territory is a mess because the Territory is a mess because the Territory is subject to all this rubbish that is fed to us from those people down there (Paul Everingham, *NT Parliamentary Record* Part 1, 7 June 1984:517).

Such was the legislative environment surrounding the creation of the new Town Council, a climate of intergovernmental tension which had been building up over several years of mounting frustration in the Territory over the lack of progress in fully developing the mining potential of the Uranium Province. The decision to allow uranium mining at Roxby Downs in South Australia but to deny the same conditions to the much earlier discoveries at Jabiluka and Koongarra particularly rankled with the local mining industry and government *and* Opposition politicians in the Territory.

In the final section of this chapter we examine some of these inter and intragovernmental relationships and their practical effects on developments in the Uranium Province.

Inter and intragovernmental relations over Jabiru

> On a bitter winter's day, two porcupines moved together to keep warm, soon hurt each other with their quills, so they moved apart, only to find themselves freezing again. The poor porcupines moved back and forth freezing and hurting until they finally found the optimum distance at which they could huddle together in warmth and yet not pain each other too much ...(Parable, after Rhodes, 1980:270).

There are a number of 'porcupines' of a governmental kind involved in the Uranium Province and by far the largest of them are the commonwealth and Territory administrations. It will be recalled from chapter 1 that we identified four broad areas of governmental conflict, following Saunders (1984), corresponding to organisational, economic, political, and ideological dimensions. They are advanced as a means of bringing some order to the complex mass of policy decisions by the various bodies concerned with Jabiru and each is now examined drawing upon examples of issues requiring resolution during the establishment of the town. The increasingly open conflict between the Hawke Labor government and CLP administration in the Territory is only the most obvious area of tension but there were also others at different levels of government—the relationship for example between the Development Authority and Town Council and between both of these bodies and Territory government agencies.

Some idea of the complexity of the situation in Jabiru in mid-1983 can be gauged from Figure 4.3, where it can be seen that three broad tiers of governmental and quasi-governmental agencies and bodies are involved. No attempt is made here to deal with these interrelationships in their entirety, the objective being to illustrate major dimensions of conflict using the fourfold typology. It is also clear that this approach does not attempt to deal with relations between governments and corporations participating in giant resource projects as these are covered by well established theories such as pluralist, corporatist and Marxist ruling class models (Galligan, 1983).

Organisational tensions

These are defined by Saunders (1984:89) as the 'different institutional interests which arise out of the different degrees and types of resources controlled by different government agencies' at the central and sub-central levels of government. In particular the sort of issues which arise when there is conflict between 'central direction and sub-central autonomy and self-determination'. There is a peculiar division of responsibility between governments in the Uranium Province with a high degree of federal control exercised through the National Parks Service rather than a conventional ministry and local bodies which are politically accountable in the Territory. There is added complexity because the Parks Service sees its role as looking after the interests of Aborigines in particular, whereas those of other local residents are handled by Territory agencies and political institutions. A good example of the tensions involved was seen in the failure of commercial interests in Jabiru in 1983 to secure a licence to sell take-away liquor.

The basis of the dispute was more than a simple conflict between Jabiru residents and the Director of the Parks Service who has ultimate responsibility for developments in the town, but was a clash of interest between the differing priorities of a largely white mining community who were in favour of 'normal' conditions for the sale of liquor, and local Aboriginal groups who were not. The original application by people in the town to acquire a licence from the Northern Territory Liquor Commission rapidly escalated into a confrontation between the Parks Service and the NLC on one side and the Territory

Figure 4.3 Chief bodies concerned with the planning, development and management of Jabiru, NT, 1983

	ADVISORY BODIES	GOVERNMENT/EXECUTIVE BODIES	OTHER
COMMONWEALTH/ NATIONAL	O.S.S. / U.A.C. / A.I.A.S.	Home Affairs and Environment / Trade and Resources / Aboriginal Affairs — Director A.N.P.W.S.	A.M.I.C. / A.A.E.C. / A.F.C.
STATE/ REGIONAL	S.I.U.M.P. — A.R.C.C. / A.R.R.R.I.	D.A.A. / W.A.C.C. — N.T. Chief Minister / N.T. Law / N.T. Mines and Energy / N.T. Transport and Works / Other N.T. Departments / Kakadu N.P. — J.T.D.A.	N.L.C. / S.C.S.I.N.T.
TOWN/ LOCAL	J.T.A.C.	J.A.W.G.	E.R.A. / Pancontinental / Denison / Gagudju Association

Definitions:

AAEC	Australian Atomic Energy Commission	JAWG	Jabiru Area Working Group
ACF	Australian Conservation Foundation	JTAC	Jabiru Town Advisory Council
AIAS	Australian Institute of Aboriginal Studies	JTDA	Jabiru Town Development Authority
AMIC	Australian Mining Industry Council	NLC	Northern Land Council
ANPWS	Australian National Parks & Wildlife Service	OSS	Office of the Supervising Scientist
ARCC	Alligator Rivers Co-ordinating Committee	SCSINT	Standing Committee on Social Impact of Uranium Mining
ARRI	Alligator Rivers Region Research Institute	SIUMP	Social Impact of Uranium Mining Project, AIAS

government and many residents on the other. At issue was a constitutional wrangle about the validity of Northern Territory law or commonwealth Regulations under the Kakadu National Park's Plan of Management (ANPWS, 1980). The outcome was decided in favour of the commonwealth by the creation of new Regulations which determined that all future liquor licence applications must be approved in the first instance by the Director of the Parks Service in consultation with the NLC (*The Australian* 20 August 1983).

Although it is usual for commonwealth law to take precedence in situations of this kind, the case caused considerable resentment in Darwin and demonstrated the inability of Territory agencies to operate in the Uranium Province if their actions were perceived to be against the policies of the commonwealth or NLC. It was almost as though a 'state within a state' had been created:

> This indecent haste makes nonsense of your hope that sensible co-operation will prevail at Jabiru as it has in the past. The gazetting of these regulations is a further example of the unwillingness to consult with the Northern Territory Government (Telex from P. A. E. Everingham, Chief Minister, to R. J. L. Hawke, Prime Minister (*The Australian* 20 August 1983).

Economic conflicts

Reference has been made to the way in which commonwealth and Territory governments prevailed on the mining companies to pay for the headworks and most social infrastructure in Jabiru, but there were also intergovernmental disputes about the provision of other facilities in the town. The most important of these relates to the perennial shortage of private sector (non-mining company) housing in mining towns which was exacerbated in this case by several additional factors.

The second Chairman of the Authority, Geoff Stolz, realised that a serious shortage of housing for commercial interests was looming by mid-1981 and sought the advice of the Territory Treasury. The response suggested that there were two prime areas of need: for established businesses such as banks requiring 'sucessor in office' accommodation for employees, and for varying numbers of service personnel working in the town. Needs of the former were to be met by the businesses concerned but the latter group were really the responsibility of the Northern Territory Housing Commission. But Jabiru was legally a 'closed' town and the Northern Territory government was loath to spend housing funds in a place where it had so little formal presence. It took the position that the commonwealth could fund this and other aspects of the social infrastructure which were not being paid for by the companies until such time as it was prepared to hand over control of the uranium mines to the Territory. In the words of the Minister for Lands and Housing, 'There would have to be a change in NT government policy before we would allow the Housing Commission to move into Jabiru and build up a stock of general public housing' (*NT Parliamentary Record* Part 2, 13 October 1982:1017). There was also the additional problem that the high standards demanded for dwelling construction in the town and the expense of site establishment would lead to very costly public housing indeed. The matter is still unresolved but part of the pressure has been eased by the retention and renewal of some of the 'temporary' accommodation in the old mine camp at Jabiru East. The commonwealth, for its part, has shown no willingness to provide special capital funding for the purpose though it does make certain grants to the Territory in overall recognition that it withholds control over uranium mining royalties.

Some 140 vacant and serviced lots remain in the town as a result of the failure of Pancontinental and Denison to commence their mining operations (Map 13). They were financed by the Authority using commonwealth loan funds and their potential sale price has risen sharply with accrued interest on the borrowed capital. The question has now arisen whether this land will be kept for Pancontinental's possible future use (Denison has decided to operate a fly-in fly-out scheme from Darwin instead of basing its workforce in Jabiru) or used for public housing. What appears most likely is that part of it will be sold to the private sector when and if tourism-led diversification is allowed to take place and locate in the town.

The political dimension

Saunders (1984) defines this as the tensions which occur between the objectives of corporate agencies with their emphasis on efficient strategies and the demands of democratic accountability placed on elected government. In the early years of development in the Uranium Province the dominant interests in both public and private sectors were the highly organised agencies of commonwealth and Territory governments with their substantial budgets and resources, together with equally well organised and independent corporate mining capital and minor pastoral interests. The small group of diffuse and largely unheard local Aboriginal people represented the only open mode of interest articulation.

Later on, the formation of the Gagudju Association in 1979 and the Jabiru Town Advisory Council in 1982 introduced open and often contradictory, pluralist, pressures upon government agencies such as the National Parks Service and the Northern Territory Department of Health. The introduction of some local democracy in the Uranium

Province strengthened pluralist interests in a manner which does not at first sight fit easily into the context of a strong and interventionist Territory government. It was part of an overall strategy, however, to create the pre-conditions for decentralised development in the Territory after the post-cyclone Tracy building boom in Darwin began to wane:

> Just as the building boom was about to go bust, I had the privilege of leading the Northern Territory to self-government. With self-government, the Territory embarked on a deliberate policy of decentralisation, though it was a matter of showing people where the opportunities were and *devolving power* along with the funds to make things happen outside Darwin. New towns were created. Jabiru is the capital of the Territory's uranium province, and shows signs of eventually becoming the major tourist centre serving the Kakadu National Park... The Northern Territory Government gave these communities the power, the money and the responsibility, and they got on with the job.
>
> The key to people-led decentralisation lies not with big government working out of capital cities, but with the third tier of government—city and shire councils (Paul Everingham in *The Australian* 17 January 1985) (emphasis added).

Territory involvement in quickening the pace of local government introduction in Jabiru was also a means of exerting maximum leverage in a resource-rich area dominated by outside public and private corporate bodies. In this sense local government in Jabiru forms part of a long-term strategy to wrest control from commonwealth and mining companies and to maximise growth potential in the regional interests of the Territory. It was seen as the vehicle to achieve 'open town' status in Nhulunbuy and Jabiru and the eventual diversification of their economies. Whether such a change is beneficial to all those who live in and around the two towns is not as clear cut and there is a continuing need to view these settlements as part of their sub-regions rather than as 'white' enclaves surrounded by Aboriginal land (Lea, 1984).

An ideological dimension

This last area of conflict covers the tensions arising between the rights of private property and those of citizenship or, put another way, between individualism and collectivism (Saunders, 1984). This ideological distinction is not obvious in Jabiru itself where all land is owned by the commonwealth and where a limited term sub-lease marks the fullest extent of private ownership. It is a situation which immediately highlights the crucial absence of private freehold property ownership in comparison to almost all other local government areas in Australia and must be contrasted with the 'normalised' mining towns of the Pilbara. Aboriginal rights over land surrounding the town are in a different category of ownership but restricted by the 99-year lease to the commonwealth for use as a National Park. Thus local residents, both black and white, face considerable barriers to investment and development in land.

This situation works quite well under the circumstances of a 'closed' mining community but experiences increasing contradictions as the town becomes more involved in tourism related functions. It is hard to imagine, for example, how significant private investment in tourist facilities can be attracted to the region without greater security of tenure than is presently available. Part of the answer is the increasing financial involvement of the Gagudju Association and it is likely that they will be the first to gain permission to build a hotel in the town (*NT News* 2 May 1985). Conflict could also arise over future attempts to enlarge the town site, a problem which has already occurred in Nhulunbuy and has not been addressed there because of fears by the mining company involved (NABALCO) that such action might lead to a review of its overall agreements with the commonwealth. The latter includes extremely favourable royalty conditions when compared with later mining agreements in the Territory (Altman, 1983). There is greater provision for future expansion at Jabiru although the environment constraints attaching to the National Park will probably restrict its development to all but future mining and tourism activities.

We have illustrated a sample of governmental tensions which emerged during the development of Jabiru but the question remains whether the two bodies with the greatest influence on its future, the commonwealth and Territory governments, have found the 'optimum distance' between them to use the words of our opening parable. There is some evidence they have now reached an accommodation over the future of the Uranium Province and that the process of conflict and withdrawal is nearing an end. The new plan of management for the enlarged National Park will need to address the suitability of Jabiru for tourism functions and there are signs that the involvement of Aborigines in such investment has provided the pre-conditions for this change in policy to occur.

A recognition by the Territory government of the *fait accompli* over further uranium mining for the time being can be seen in the winding down of the Development Authority and the increasing attention being given to tourism rather than mining as the mainstay of the Territory economy. Kakadu remains as the prime attraction in the Top End outside Darwin and the argument is now about how to maximise that potential.

We turn now in Part III to an analysis of development and change in Jabiru from the perspective of the residents themselves. To what extent, for example, is there any evidence of widespread popular concern over town administration or restrictions over lifestyles caused through the location in a National Park? Who are the people who live in Jabiru and how do they make use of their leisure time, and so on. We hoped to discover whether the residents' views supported the evidence obtained from interviews with Development Authority officials and community leaders and to judge if there is widespread apathy about town affairs among the great majority of residents who do not normally air their views or vote in elections.

Part III   LIVING IN JABIRU

5 A community profile

Now that I have been here awhile I have begun to think of Jabiru as home,
Agree: 71%. (Zehner and Lea, 1984:49).

A profile of the residents of Jabiru provides a base from which to
examine a variety of responses to life in the new mining community
and to changes in governance arrangements which have taken place
since 1982. We indicated in chapter 1 that sociologists have attempted
to describe the Western European mining town in terms of some eight
ideal types or generalisations (Bulmer, 1975). Although we will
consider the applicability to Jabiru of these characterisations in the
concluding chapter, it seems likely that Australian resource towns
could exhibit their own distinctive social profiles, some aspects of
which may differ considerably from the European stereotype.

In this chapter we analyse data from the July 1984 Community
Survey (Zehner and Lea, 1984) to shed light on Jabiru's social
characteristics and on the residents' experience of living there. In
particular we examine reasons why people came to live in Jabiru and
how they subsequently evaluated their housing and the quality of the
town's facilities and services. In addition to comparisons of married
Ranger and non-Ranger households, the responses of single persons in
Jabiru, who form some 40 per cent of the households, are examined
separately.

Jabiru at the time of the community survey

The six to eight months before July 1984 were a period of considerable
activity in Jabiru as the chronology in chapter 4 made clear. We found
on a visit to the town early in 1984 that two of the top management at
Ranger, the General Manager and the Manager of Operations, had
only recently left the company's employment. Although retirement
was given as a reason for their departure, there was widespread
suspicion in Jabiru that they had been retired because ERA did not

97

think that they would be tough enough in dealing with union demands for better pay and conditions. It was a circumstance which appeared to follow the pattern of changing management observed in other new mining towns once the production phase of mining operations has been established. There had been labour unrest in Jabiru in November 1983 which included eight strikers being arrested for lying down in front of a staff-driven truck owned by ERA (*The Australian* 21 November 1983). Forty members of the two unions involved, the Federated Miscellaneous Workers Union and the Australian Society of Engineers, had formed a 24-hour picket line across the main road between Jabiru and the mine.

The company stood firm against the four main issues in the dispute: the introduction of a redundancy payment scheme, a superannuation scheme, a relocation allowance for employees terminating their employment and a cost of living increase (Notice to all award employees from ERA Chief Executive, 13 December 1983). ERA took the view that these demands contravened the terms of Principle 11 of the National Wage Case at the time. They had decided, in addition, to unilaterally and substantially raise the rents on company housing and in November 1983 the weekly rent for a Ranger house in Jabiru was

Picket line on the road to Jabiru East, November 1983.

increased from $15.50 to $65.00. Ranger wage employees were infuriated (see Figure 5.1).

In a textbook example of the power that can be wielded by the mining company in a company town, ERA agreed to waive the rent increase if employees returned to work. The dispute was not a short one, however, and it brought sharp social divisions in the town to the surface before it was finally settled in January 1984. ERA had not given ground on the four key work conditions at issue and the consensus in Jabiru was that the unions had caved in. Although ERA had 'won', the Ranger management was left tired and unsettled by the events of the previous months. The unions had 'lost' most of their demands and while rents did return to pre-strike levels, their members had little to show for their time off except lower bank balances. In the words of the wife of one Ranger (wages) employee in the July survey: 'The strike really messed the town up. [Before the strike Jabiru] was a good friendly community [but] post-strike the ill feeling was very strong.'

We were concerned as researchers that the turmoil surrounding the strike and its clear winner-loser conclusion would make it difficult for anyone to conduct a survey of residents as soon as July, but our academic commitments provided little leeway for postponing the fieldwork. The decision was made, however, to avoid questions in the interview that could be interpreted as attempts to reopen or exacerbate industrial problems in the community. In part we wished to see how often comments concerning industrial strife would arise spontaneously in response to questions about 'community issues', and we also certainly did not want to antagonise one or another faction in the town. Our focus, after all, was on community governance and *all* residents rather than on union and mine relationships. Obviously any attempt on our part to draw generalised conclusions about Jabiru would have been seriously undermined had sections of the community systematically refused to take part.

Fortunately, most residents who fell into the sample chose to respond and the impression we received was that people rather enjoyed the opportunity to talk about the pros and cons of living in Jabiru. Interviews were obtained from 89 per cent of the occupied households sampled, a very good rate of response for a household interview survey in the 1980s, and notably high given the difficulties of contacting and interviewing single respondents in mining towns. (See Appendix B for a review of the study's sampling design and response rates.) In all, 323 interviews were obtained from a sample which included 364 of the 607 dwellings in Jabiru occupied at the beginning of July 1984. From these interviews, 118 of the households were classified as Ranger married, 100 as non-Ranger married, 64 as Ranger single and 41 as non-Ranger single.

FOR PUBLIC

INFORMATION

The following resolutions were adopted at a combined union meeting on 25/11/83 an presented to Ranger Management.

That a motion of scathing condemnation be expressed against Ranger for its action in increasing Rental and Accomodation charges by over 400% while it expects us to be satisfied with wage guideline increases of 4.3% .

That Ranger Management immeadiately cease using Staff labour to operate the mine, and that they will be held responsible for creating extreme social divisions in the town of Jabiru by their actions.

Figure 5.1 Union protest notice displayed during the strike in Jabiru, November 1983

It is important to recognise that categorising respondents into these groupings was not as simple as might first appear. Problems of classification occur, for example, when one spouse works fulltime for Ranger and the other fulltime for someone else (or for themselves); when a married individual lives in single persons quarters; or when a single parent with his or her children lives in housing customarily allocated to married couples. Overall, almost 20 per cent of the households interviewed did not fit neatly into one of the four employer/marital status categories.

In the tables which follow respondents living in households where both spouses were employed fulltime (one with Ranger and one not with Ranger) were classified as 'Ranger marrieds', if they lived in Ranger housing, or as 'non-Ranger married' if they did not live in Ranger housing. Single parents were classified as single even if living with their children, and married individuals living alone (or living in shared accommodation with other non-marrieds) were also considered to be single. Finally, respondents who reported being in a de facto relationship were classified as married.

A community profile

One of the most apparent characteristics of new planned communities is the definition of 'oldtimers' as people who arrived only two or three years before. In this sense Jabiru is largely a settlement of newcomers. Only a handful of the present residents had lived in the Jabiru area or even the Alligator Rivers Region prior to 1979.

Table 5.1 shows that Ranger married households, as a group, are the most established. Sixty-three per cent had arrived by 1981, the year production at the plant started. Equally interesting is that almost half (48 per cent) of the single non-Ranger respondents arrived in 1984 in the six months preceding our survey. Almost a third (31 per cent) of those interviewed moved to Jabiru from elsewhere in the Territory and another 21 per cent arrived from Queensland. Non-Ranger households were more likely to have come from elsewhere in the Territory; singles were more likely to have come from Western Australia (mainly the Pilbara) than marrieds; and marrieds were slightly more likely to have come from Victoria than singles.

A majority of the Ranger respondents had lived in at least one other mining town before Jabiru. Most non-Ranger people had not, and for 71 per cent of the married non-Ranger respondents Jabiru was their first mining town. The interview also asked people in which other mining towns they had lived. The most frequent mentions were of places in northern Australia or nearby Papua New Guinea: Mt Isa,

Table 5.1 The people in Jabiru
(percentage distributions of respondents)

	Ranger married	Non-Ranger married	Ranger single	Non-Ranger single	Total
Year of arrival					
1980 or earlier	21	17	14	7	16
1981	42	13	25	11	25
1982	16	15	14	14	15
1983	9	31	11	20	16
1984	12	24	36	48	28
Previous residence					
Northern Territory	25	43	27	34	31
Queensland	26	19	17	19	21
New South Wales/ACT	9	14	14	19	13
Western Australia	8	2	16	14	10
South Australia	11	8	8	6	9
Victoria	10	5	4	1	5
Tasmania	3	3	3	—	2
Overseas	9	6	12	8	9
Number of other mining towns lived in					
None	43	71	46	61	53
One	30	21	24	13	23
Two	13	5	11	8	10
Three or more	14	3	19	18	14
Marital status of respondents					
Married	87	87	12	6	54
De facto	13	13	—	6	9
Single	—	—	67	80	31
Separated, divorced, widowed	—	—	21	8	6
Number of respondents[a]	118	100	64	41	323

Note: a Percentages are based on the number of people responding to each item. In most instances responses are available for all (or almost all) items from virtually all respondents. Situations where the number of people responding varies greatly from that noted here will be indicated in later tables where appropriate. Jabiru total percentages have been calculated using data weighted to reflect the survey sampling design. See Appendix B

Qld. (18 respondents); Tennant Creek, NT (14); Bougainville, PNG (11); and Nhulunbuy, NT (9), although well over 100 other mining communities elsewhere in Australia and around the world were mentioned by one or more respondents.

The final entries in Table 5.1 concern marital status. Sixty-three per cent of the respondents described themselves as married or in a de facto relationship. Of particular interest is the finding that roughly one-eighth of the people living as singles in Jabiru were actually married and either planning to rejoin their spouses after a period of employment in the town or waiting for their families to arrive.

A large amount of information about Jabiru households is detailed in Table 5.2. The first entries indicate a marked similarity across the comparison categories in the median ages of male and female heads of household in the town. We had expected the single population to have been younger than their married counterparts, but the differences were minor in the case of the men and virtually non-existent in the case of women. Interestingly, the median ages for male (34 years) and female (32 years) household heads in Jabiru are almost the same as those of male and female adults (age 20 and over) in the Territory as a whole. According to 1981 Census data the median age for Territory men at that time was 33; for women it was 32 (ABS *Summary Characteristics of Persons and Dwellings Northern Territory* 1982: 6,7). Age data presented in aggregated form for six mining towns in the Pilbara also suggest similar distributions to those found in Jabiru with median ages in those communities apparently near or slightly above 30 (Neil et al., 1982: 43).

The next panel in the table indicates that married Ranger households were considerably more likely to have four or more members than were the non-Ranger married households. Also of

Table 5.2 The people in Jabiru — characteristics of heads of households
(percentage distributions)

	Ranger married	Non-Ranger married	Ranger single	Non-Ranger single	Total
Age					
Male head					
Under 25	8	5	34	28	18
25-34	41	48	33	37	40
35-44	34	33	19	26	29
45-54	10	11	8	9	9
55 and over	7	3	6	—	4
Median age	35	34	32	30	34
Female head					
Under 25	16	18	25	32	18
25-34	43	52	38	34	45
35-44	28	19	37	14	25
45-54	9	8	—	15	8
55 and over	4	3	—	5	4
Median age	32	30	33	30	32
Household size					
One person	—	—	93	91	41
Two	27	31	3	5	17
Three	17	30	1	2	12
Four or more	56	39	3	2	30

Table 5.2 (Continued)

	Ranger married	Non-Ranger married	Ranger single	Non-Ranger single	Total
Country of birth					
Male head					
Australia	62	64	65	79	65
Australian Aborigine	1	2	2	3	2
United Kingdom	16	10	9	6	12
Other Europe	9	14	13	3	11
Asia/South East Asia	2	1	6	3	2
Other	10	7	5	6	8
Female head					
Australia	65	66	90	62	67
Australian Aborigine	2	3	—	12	3
United Kingdom	15	9	—	13	12
Other Europe	5	6	—	—	5
Asia/South East Asia	4	5	10	—	4
Other	9	11	—	13	9
Education					
Male head					
Up to 10 years	49	33	35	24	39
11-12 years	21	27	22	24	23
13 or more years	30	40	43	52	38
Female head					
Up to 10 years	51	39	70	38	46
11-12 years	30	30	30	25	30
13 or more years	19	31	—	37	24
Employment status					
Male head					
Employed fulltime	100	99	100	97	99
Employed part-time	—	1	—	3	1
Female head					
Employed fulltime	30	28	100	87	34
Employed part-time	30	33	—	—	29
Unemployed	3	3	—	—	3
Household duties	38	35	—	13	34
Employer (of those employed)					
Male head					
Ranger	99	—	100	—	58
Ranger subcontractor	—	10	—	18	5
OSS	—	22	—	18	9
NT Government	—	27	—	3	9
Private sector	1	28	—	43	14
Town Council/JTDA	—	7	—	3	2
Other	—	6	—	15	3

Table 5.2 (Continued)

	Ranger married	Non-Ranger married	Ranger single	Non-Ranger single	Total
Female head					
Ranger	28	13	100	14	26
Ranger subcontractor	7	7	—	—.	7
OSS	2	4	—	14	3
NT Government	15	30	—	29	20
Private sector	34	26	—	29	29
Town Council/JTDA	5	9	—	—	6
Other	9	11	—	14	9
Occupation category					
Male head					
Professional, managerial, administrative	21	62	26	39	37
Clerical, sales	5	4	6	—	4
Skilled trade	33	15	34	37	28
Unskilled trade	41	19	34	24	31
Female head					
Professional, managerial, administrative	22	34	20	58	28
Clerical, sales	31	18	20	14	24
Skilled trade	4	3	20	14	5
Unskilled trade	43	45	40	14	43
Level of supervision					
Male head					
Non-supervisory	63	33	77	53	55
Supervisory	27	41	19	25	30
Managerial	10	26	4	22	15
Female head					
Non-supervisory	76	73	87	57	74
Supervisory	15	17	13	—	15
Managerial	9	10	—	43	11
Family income					
Under $10 000	1	—	12	6	5
$10 000-$19 999	10	13	19	30	17
$20 000-$29 999	37	35	46	36	38
$30 000-$39 999	34	32	16	22	27
$40 000 and over	18	20	6	6	13
Number of households including					
Male head	118	100	54	33	305
Female head	118	100	10	8	236

interest is that small proportions of the single households included more than one person. This occurred when accommodation was being shared with other singles or because the respondent was a single person with children.

An overwhelming majority of both the male heads (67 per cent) and female heads (70 per cent) in Jabiru were born in Australia, although only a few were Aboriginal Australians. The next most frequent birthplace was the United Kingdom (12 per cent for both men and women). The percentages for Australian-born are higher, but not too dissimilar to those found by Stockbridge et al. (1976: 54) in the Roebourne Shire in the Pilbara (63 per cent Australian-born; 24 per cent born in the United Kingdom) and with those reported by Neil et al. (1982: 45) in their review of 1981 data from Paraburdoo in the Pilbara (males: 64 per cent born in Australia and 20 per cent in the United Kingdom; females: 71 per cent Australia and 13 per cent United Kingdom). The five other mining-based towns included in Neil's review all had somewhat lower percentages of Australian-born in their populations (down to around 50 per cent in Newman in 1980) and correspondingly higher numbers from the United Kingdom (around 27 per cent in Newman in 1980).

Asking people to quantify the amount of formal education they have attained (or endured) is not a straightforward task like asking about age or country of birth, particularly when a number of the respondents were overseas for much and possibly all of their schooling. Further, even within Australia standards and labels for different levels of achievement vary as well. Given these difficulties we chose to focus on a 'simple' measure of the number of years individuals had attended school. This clearly does not eliminate the problem of measuring education, but the summary of education levels does suggest that singles tended to have more formal education than the marrieds; non-Ranger people more than those in Ranger households; and men more than women.

The employment situation for heads of household is summarised in the next panels of the table. Virtually all of the male household heads were employed on a fulltime basis. Roughly one-third (34 per cent) of the female heads of household were also employed fulltime and an additional 29 per cent were occupied for varying amounts of time from a token few hours per week to virtual fulltime employment. None of the male heads and only three per cent of the female heads were classified as unemployed, a situation which helps emphasise one of the differences between mining towns and 'normal' communities in Australia.

Predictably, the employed male heads in Ranger married and Ranger single households were mine employees (with only one exception where the Ranger employee in the house was the wife rather than the husband). There were three main sources of employment for the non-Ranger households: the Northern Territory government (for example, police, teachers, health care), the Commonwealth Office of the Supervising Scientist, and the private sector. Despite the obvious dominance of Ranger in Jabiru it is notable that 42 per cent of the male heads and 74 per cent of the employed female heads did *not* work for Ranger. In fact, if one adds up the total number of pay cheques implied by these figures, over half would come from a source *other than* Ranger, albeit often for part-time employment.

The most striking aspect of the next panel of results is the higher proportion of professional/managerial/administrative positions among non-Ranger compared to Ranger employees. Most Ranger jobs are in skilled or unskilled trades occupations, especially among men, a fact that is reflected in the relatively smaller number of Ranger employees holding supervisory or managerial positions compared to those in the non-Ranger workforce.

Compared to the men's, the women's occupations included a much higher percentage of clerical/sales jobs. As a group the women were much more likely than the men to work in non-supervisory positions, although three of the seven single non-Ranger women for whom we have data did report having managerial occupations.

The final part of Table 5.2 concerns family income. Fifty-nine per cent of Ranger married and 62 per cent of non-Ranger married households had two incomes which largely accounts for the significantly higher married incomes compared to those of the singles (there was a predictably strong correlation, $r = .39$, between family income and the number of employed heads in a household). Based on 1981 Census data, adjusted to allow for an annual increase of eight per cent, the median income for Northern Territory families at the time of our survey would have been approximately $18 765 (ABS *Summary Characteristics of Persons and Dwellings Northern Territory* 1982: 16). For Jabiru the comparable figure was $27 377. In aggregate, any community with 50 per cent of its households earning that much (or more) clearly has a number of affluent families. But of equal interest is that over one-fifth of the households earned less than $20 000 in the preceding year and five per cent reported incomes under $10 000 (sometimes due to prolonged unemployment before they moved to Jabiru). For many of these people, particularly the married households with incomes under $20 000, the notion that 'anyone working in a mining town must be doing very well financially' would have appeared more illusion than reality.

This section of chapter 5 has provided a profile of the residents in

Jabiru as of July 1984, a point in time when the mining operation was solidly into its production phase. In brief, the survey data found that a 'typical household' had been in Jabiru for only two or three years, was young, married (or paired in a de facto relationship) with both husband and wife born in Australia and living on a healthy income with one or more persons employed by Ranger.

Searching for 'modal' households glosses over a number of contrasts, however, the most important being the presence of a large number of single households and of households not directly employed by Ranger. Other than the obvious difference in their employers, one of the most striking of the contrasts between Ranger and non-Ranger households in Tables 5.1 and 5.2 was the difference in education levels. Heads in non-Ranger households had completed more years of formal education. Non-Ranger families were also, in aggregate, the more recent arrivals in Jabiru and were less likely to have lived in another mining town prior to coming to the community. On the other hand, non-Ranger residents were *more* likely to have come to Jabiru from somewhere else in the Northern Territory and can be assumed to have had some idea of what living in another town in the Territory would be like.

The following section focuses on the reasons people gave for moving to Jabiru and on comparisons of Jabiru with their previous residence and, where applicable, with other mining towns in which they had lived.

Moving to Jabiru

In a metropolitan area questions about why a community was chosen as a place to live tend to elicit a variety of responses ranging from the quality of nearby schools and the cost and layout of housing to thinly veiled comments about social status and 'desirability' of neighbourhoods (for example, see Lansing et al., 1970; Kilmartin and Thorns, 1978; Stimson, 1982). Convenience to work is often an important consideration as well, but one which would only limit choice to a range of different communities within, say, 30 or 45 minutes of the workplace. For most people moving to Jabiru, however, there were few concerns about subtle differences in housing type and community amenity. The overwhelming reason, and often the only reason, mentioned for coming to Jabiru was, in the words of a research scientist with the OSS, 'Because this is where the job was.'

Finding work, sometimes after a period of unemployment or after another mine closed down (as with Gunpowder or Mary Kathleen in Queensland), was given as a reason for coming to Jabiru by 74 per

cent of the Ranger married respondents and up to 90 per cent of the
non-Ranger single respondents (see Table 5.3). Since Jabiru is a closed
community in that people who are not directly or indirectly involved
in the mine (or in supporting the community) cannot obtain housing,
the predominance of work-related reasons for coming to Jabiru is not
a surprise. Interestingly, although all respondents were asked if there
was more than one reason for their decision to come to Jabiru, half of
them were unable to come up with anything beyond the job. Further,
while the attraction of Kakadu and the environment around the town
was the second most often mentioned reason for coming to Jabiru
(mentioned by only 23 per cent overall), a general response which
arose next most frequently, 'to make some money' (mentioned by 18

Table 5.3 The move to Jabiru
(percentages of respondents)

	Ranger married	Non-Ranger married	Ranger single	Non-Ranger single	Total
Reasons for moving to Jabiru[a]					
Work; job for spouse	74	87	83	90	82
To make money	23	11	21	13	18
Environment; Kakadu NP	27	18	11	30	23
Friend or relative in town; social reasons	16	11	22	5	14
Jabiru compared to last place of residence					
Better	51	52	43	42	47
About the same	21	17	11	24	19
Not as good	27	30	41	30	31
Don't know	1	1	5	4	2
Jabiru is 'better' because					
Relaxed life style	34	39	26	23	31
Friendly people; good community spirit	28	38	25	33	30
Environment; Kakadu NP	26	29	29	29	29
Good facilities; nice community n.e.c.	26	18	4	8	15
Jabiru is 'not as good' because					
Not a friendly town; no spirit; divisions in town	24	25	34	37	30
Facilities lacking; too small a town	23	23	42	27	29

Note: a Up to three responses were coded for these open-ended questions so that the
figures may add up to more than 100 per cent. To focus attention on the issues
which were of most interest to the respondents, infrequent 'other' responses
have been omitted from this and subsequent tables.

per cent), would also have been closely tied to obtaining work in Jabiru. CSIRO surveys of Pilbara mining communities have found similar types of reasons for moving into mining towns and 'employment' and 'finance' were mentioned by 68 per cent (Hedland) to 93 per cent (Shay Gap) of their respondents (Neil et al., 1982: 37).

Jabiru residents were also asked to compare Jabiru with the last place that they lived. This proved to be a difficult question for a number of respondents who had never lived in a mining town before and who found it hard to compare, say, Jabiru with Melbourne or Perth. Nevertheless, most people were able to decide if they thought that Jabiru was 'a better community to live in, not as good, or about the same'. While almost one-third of the respondents felt that Jabiru was 'not as good', 47 per cent felt that it was a better place to live, including one married Ranger employee who said 'For me, I reckon it's magic!' Single respondents were noticeably less likely to rate Jabiru as 'better' than were married residents.

A probe of these ratings followed in the interview. The main reasons for 'better' responses revolved around the friendliness of the community, the relaxed lifestyle that was possible in Jabiru, and the attractions of the environment and the surrounding Kakadu National Park. Curiously, the main reason given for Jabiru being 'not as good' was that the community *lacked* the friendliness and communal spirit people had found in other places. The other major category of criticism focused on the problems of Jabiru's size and the facilities and types of people (especially single women) that they felt were lacking. The responses below are representative of some of the issues that were raised.

Jabiru 'better'
Male, 28, married, from Gunpowder, Qld, two years in Jabiru

> The housing is excellent...best standard I've seen in mining towns. The facilities are excellent...a wide range of activities.

Female, 26, married, from Darwin, four years in Jabiru

> I find that my children are a lot more free to play. I know where they are. The streets are safe. They have a lot more freedom than if we lived in a big town.

Male, 30, married, from Adelaide, three years in Jabiru

> It's not as crowded. The people are good, the air is clean. The people are friendlier. There's a better community spirit.

Jabiru 'not as good'
Male, 23, single, from Brisbane, three years in Jabiru

The lifestyle is different; the problems aren't different here. You just notice them here. You lose the anonymity of the city.

Female, 38, married, from Mary Kathleen, Qld, one year in Jabiru

Jabiru is more like a ghost town than Mary K. There was a better community spirit at Mary K.

Male, 31, single, from Charleville, Qld, four months in Jabiru

(Why?) The social set from my point of view. Jabiru tends to be young marrieds. Few people to meet if you aren't married. A closed town...stifling for singles.

Although not detailed here, the variables included in Table 5.3 were also tabulated against the sex of the respondent and the length of the respondent's residence in Jabiru. The only difference of note was that longer term residents were somewhat more likely to rate Jabiru as a 'better' place than more recent arrivals (52 per cent 'better' for those arriving before 1982 compared, at the other extreme, to 42 per cent 'better' for those arriving in 1984). Such a finding is not entirely unexpected since those who arrived before 1982 who found Jabiru disappointing would be more likely to have moved out—leaving the more satisfied members of their cohort in the community.

Housing evaluation

Mining towns need to provide a high standard of accommodation if companies hope to attract and retain employees. Although exceptions do exist like the tightly planned medium density housing in Goldsworthy Iron's Shay Gap, WA, the norm has been for most married housing to replicate the homes found in a middle class, low density, capital city suburb.

In the case of Jabiru, planners clearly adhered to the notion of creating a series of high quality 'suburban' subdivisions, and substantial brick homes and townhouses were constructed for the married Ranger employees. The mine, working through the Development Authority, was also responsible for the construction of a number of single persons quarters (SPQs). Accommodation for the public sector was to be provided by the commonwealth and Northern Territory governments, while housing for the private sector was the responsibility of the Development Authority (using Ranger funds). Because of the leasehold status of the townsite, Ranger housing as well as government housing was to be rented, and no attempts were made to provide a homeownership scheme for Ranger employees similar to those adopted by Hamersley Iron and Mt Newman Mining in the Pilbara, and Utah in the Bowen Basin.

There are three main types of Ranger housing within the townsite of Jabiru: 149 three and four-bedroom detached homes; 38 two-bedroom townhouses; and 132 SPQ units. At the time of our sampling, nine houses and 21 SPQs were allocated to non-Ranger personnel including, for example, the Acting Town Manager, the minister of the Uniting Church and the manager and staff of the catering firm who ran the mess. In Table 5.4 these dwelling units are included in the Ranger columns. All Ranger dwellings in the Jabiru townsite are of brick construction and are fully airconditioned. Following the recommendations of A.A. Heath & Partners (1978: 20), government housing in Jabiru was not allocated to a single part of the town and is scattered more or less randomly around Jabiru's neighbourhoods (see Map 13). It is also built in brick and of such a high standard that a visitor would be unlikely to be able to tell the Ranger and government units apart. Three-bedroom government houses, for example, provide 123 m^2 of indoor space compared to 126 m^2 in three-bedroom Ranger houses (personal communication: Graham Pattle, Jabiru Town Council, 1985). Differences are apparent, however, once one realises that none of the government homes are fully airconditioned and are, therefore, the houses that depend on flow-through louvres for ventilation and cooling.

It is worth nothing that the treatment of government housing in Jabiru is radically different from that in Nhulunbuy. Planning for that Northern Territory community preceded Jabiru's by about 10 years (Agius, 1983), and there public housing would not be mistaken for that of the company, partly because the dwellings themselves appear less substantial, and partly because most government housing has been sited in a single area, the neighbourhood farthest from the town centre.

Single persons quarters for Ranger employees include a bath and an L-shaped room that allows occupants a degree of separation between living and sleeping areas. Cooking facilities are not included and, as a result, the 22 six-unit SPQs have been sited within a reasonable walk to the mess. They are dispersed around the townsite rather than clustered in one or two areas in town, however, and an indication of how single residents and their neighbours evaluated that experiment appears later in this chapter.

Housing for government singles is dispersed in a similar fashion but differs from Ranger SPQs in several respects, the main one being that they include cooking facilities and have over twice the area of Ranger SPQs, 60 m^2 compared to 24 m^2 (personal communication: Graham Pattle, 1985). Government SPQs have three to five units per building and appear more like a single family home than the Ranger SPQs. This is partly because they have been designed with glass sliding doors which open onto a small patio at the front of the unit. Table 5.4 shows that both indoor and outdoor privacy are a particular problem in government SPQs, and this seems to be a clear reflection of difficulties created by the use of floor-to-ceiling glass as the front 'wall' of the unit.

Jabiru, like most mining towns, has an assortment of accommodation that is not included in the above descriptions. The mine's construction camp in the early years of development was located in Jabiru East, a community about five km east of the present Jabiru townsite, which was expected to outlive its usefulness once the new town of Jabiru itself was complete (chapter 3). That has not happened, partly because some people prefer to live there, but mainly because of a shortage of housing in Jabiru. The housing remaining in Jabiru East is spread over a large area with more empty blocks than occupied ones. It is occupied by both Ranger and non-Ranger personnel and includes substantial high-set houses, prefabricated structures set on slabs, caravans, demountable units and several creative arrangements that have grown up around caravans.

In addition, adjacent to the townsite itself (see Map 13) are a 50-bay caravan park and 200 Nationwide demountable units used primarily to accommodate people working for Ranger subcontractors, often on short-term contract. At the time of our survey the caravan park had 36 vans, and 76 of the demountable units nearby were listed as being occupied.

A series of interview questions gave respondents the opportunity to rate several aspects of their housing. Items focused on the adequacy of indoor and outdoor space and privacy, as well as broader evaluations of dwelling design and the respondents' overall satisfaction with their accommodation. Table 5.4 sets out the responses for the eight main types of housing available in the community.

The first entries in the table concern indoor and outdoor space. Although disenchantment with the amount of space indoors was clearly highest in the government SPQs (58 per cent said there was 'not enough'), it is worth noting that the number of respondents in government SPQs was the smallest of the eight housing categories—only 12 individuals. Even so, the contrast with the Ranger SPQs is striking, particularly in view of the relative sizes of the units. The major functional difference between Ranger and government SPQs is that the government units include cooking and dining as well as sleeping and lounge areas, but these results suggest that fitting so many different activities into the same space creates problems for a number of residents in government SPQs.

Less surprising than the negative rating of government SPQs was the finding that nearly half of the residents in caravans and

Housing types *Ranger townhouses, Jabiru* (above left); *Jabiru East high-set house rebuilt on the ground with brick veneer in Jabiru* (below left); *government family house* (above).

Nationwide demountable units also felt that they lacked enough indoor space. At the other end of the continuum, 90 per cent of those in Ranger houses felt that they had enough or more than enough room indoors; and the same per cent felt that they had enough or more than enough 'outdoor space near your house which you (and members of your family) can use for your different activities'. Comparable levels of satisfaction with outdoor space came from the residents of government houses, the Nationwide demountables and the Ranger SPQs, but the most satisfied people in terms of outdoor space were those living in Jabiru East—no one there said that they did not have enough room for outdoor activities.

The finding that some 87 per cent of respondents living in demountables felt that they had enough or more than enough outdoor space nearby was a surprise. The most likely explanation is that the 'no man's land' around the demountables (and the limited resources available in the units themselves) meant that very few people wanted to spend their free time in the area, and so made few demands upon it. An additional factor may have been that at the time of our survey only 76 of the 200 demountable units were actually occupied so that there probably would not have been a sense of overcrowding outdoors in any event.

Most types of housing in Jabiru appear to give residents 'enough privacy from neighbours' when the residents are inside their dwellings.

Table 5.4 Satisfaction with housing
(percentages of respondents)

	Ranger house	Govt house	JabE house	Ranger t'house	Ranger SPQ	Govt SPQ	Caravan	Demount-able
Indoor space[a]								
More than enough	14	23	22	5	9	—	—	9
Enough	76	58	61	69	68	42	56	46
Not enough	10	19	17	26	23	58	44	46
Outdoor space[b]								
More than enough	10	20	17	17	18	8	7	13
Enough	80	68	83	48	66	67	63	75
Not enough	10	12	—	35	16	25	30	13
Indoor privacy[c]								
Enough	93	70	94	91	84	50	87	94
Not enough	7	30	6	9	16	50	13	6
Has outdoor private space[d]								
Yes	57	41	50	35	25	17	43	56
No	43	59	50	65	75	83	57	44
Dwelling is[e]								
Well designed	46	36	39	26	35	17	45	29
Average	46	30	50	39	51	50	41	59
Poorly designed	8	34	11	35	14	33	14	12
Percent preferring another type of dwelling[f]	13	42	65	65	75	56	77	81
Dwelling satisfaction[g]								
Completely satisfied	42	23	33	17	21	8	17	24
Satisfied	54	38	39	48	58	58	50	29
Neutral or dissatisfied	4	39	28	35	21	34	33	47
Number of respondents[h]	83	74	18	23	57	12	30	17

Notes: a The question was: 'On most occasions do you and others in your family have *enough* space here in your (house/unit/caravan) to do what you want without bothering the rest of the family, *more* than enough space, or not enough?' In one-person households the question was: 'On most occasions do you have enough space here in your (house/unit/caravan) to do what you want without bothering other people, *more* than enough space, or not enough?'

b The question was: 'And what about the amount of *outdoor* space near your (house/unit/caravan) which you (and members of your family) can use for your different activities—do you have more space than you need, about the right amount, or not enough outdoor space?'

c The question was: 'The amount of space people have is not always related to whether they think they have enough privacy. As far as you are concerned, do you have enough privacy from neighbours when you are *inside* your (house/unit/caravan)?'

d The question was: 'And do you have a place *outside* where you feel that you can really have privacy from your neighbours if you want it?'

e The question was: 'How well would you say the architects did when they designed this (house/unit/caravan)? Is it well designed, poorly designed, or about average?'

f The question was: 'If you had your choice would you rather live in another type of accommodation?' [If another type] 'What?'

g The question was: 'Now, overall, how do you feel about this (house/unit/caravan) as a place to live? (SHOW CARD) Which number comes closest to how satisfied or dissatisfied you feel?

Completely satisfied (7) (6) (5) (4) (3) (2) (1) Completely dissatisfied'

h Nine respondents living in various types of accommodation in the former construction camp area in Jabiru East are not included because their housing situations were not easily translatable into the categories in this table.

Only in government housing were there sizeable percentages of respondents who felt that they did not have enough privacy. In the case of government SPQs (50 per cent 'not enough'), the stylish glass front to the units and the lack of airconditioning (which meant that windows needed to be left open) contributed to privacy problems for residents. In government houses the absence of airconditioning and the need to keep louvres open for ventilation appeared to be the main reason why 30 per cent of those respondents reported that they did not have enough privacy when indoors.

The next question in the interview asked if residents had 'a place outside where you feel you can really have privacy from your neighbours'. The most positive responses came from those living in Ranger houses: 57 per cent said that they did have a private area outside; 43 per cent said that they did not. The dwelling types that appeared to have the most problems on the basis of this item were the SPQs, especially the government units where 83 per cent said that they did not have a private outdoor space nearby.

To get a better sense of the problems people faced in their housing we also asked 'How well would you say the architects did when they designed this (house/unit/caravan)? Is it well designed, poorly designed or about average?' The 'average' rating was a common response across all eight of the dwelling types, but in most cases the 'well designed' responses clearly outnumbered the 'poorly designed' answers. This was especially the case with the Ranger houses (46 per cent 'well designed' versus eight per cent 'poorly designed') and the detached houses in Jabiru East (39 per cent versus 11 per cent). In only two situations did negative ratings outnumber positive: government SPQs (17 per cent 'well designed' versus 33 percent 'poorly designed') and Ranger townhouses (26 per cent versus 35 per cent).

Probes of these ratings indicated that 'well designed' in the Ranger houses often meant having airconditioning and having the sleeping and living areas clearly separated so that shift workers did not disturb sleeping members of their families. Also appreciated was the separation of the main living room (available for use by adults) from a family area near the kitchen which often became the focus of children's activities. Negative comments regarding government houses centred on the lack of airconditioning and/or a feeling that houses which depended on louvres and flow-through ventilation should have been high-set rather than constructed at ground level, and several respondents remarked that government houses seemed to be designed for Canberra and southern capitals rather than the tropics. We have already noted the problems with privacy that individuals living in the government SPQs reported. The other dwelling type where 'poorly designed' ratings outnumbered 'well designed' was the

Ranger townhouses. Complaints covered many difficulties ranging from a lack of space for storage and visitors and problems with the heat of the summer sun to worries about being able to watch people go into and out of the upstairs bathroom from the downstairs living area.

Another indication of housing performance from the residents' perspective is the number who would prefer to be living in something other than their present accommodation. Ranger houses come out especially well on this criterion; only 13 per cent of those living in them said that they would prefer something else. Forty-two per cent of those in government houses and more than half of those in each of the other six types of dwellings said they would rather live in another type of accommodation. We asked those who disliked their present type of housing what they would prefer. Many had fairly limited changes in mind such as adding more space or amenities to the housing they already had. Some specified other housing types, which is really what we had in mind when we included the question. Specifically, about a third of those who wanted a change from government houses indicated that they would prefer a high-set house in Jabiru East; another third desired a Ranger house in town. About half of those in the Ranger townhouses who wanted to switch preferred a Ranger house, and a similar proportion of those in Ranger SPQs would have liked moving to a house or a townhouse.

The final entries in Table 5.4 turn to a summary rating of the housing, and it is clear that most residents found their housing satisfactory. By far the most favourable assessments went to the Ranger houses, with 42 per cent of respondents 'completely satisfied' and only four per cent less than satisfied. In contrast, 39 per cent of residents in government houses rated them less than satisfactory with only 23 per cent finding them completely satisfactory. Other dwelling types where a neutral or dissatisfied response decidedly outnumbered the positive response were government SPQs (34 per cent versus eight per cent), Ranger townhouses (35 per cent versus 17 per cent), demountable units (47 per cent versus 24 per cent), and caravans (33 per cent versus 17 per cent).

Most Jabiru households (65 per cent) live in the townsite in Ranger houses and SPQs and government houses and SPQs. When these groups are examined one finds that although there are some households who are *not* satisfied with Ranger housing and some in non-Ranger housing who *are* satisfied, in aggregate the results indicate a clear differentiation between Ranger and non-Ranger residents in terms of housing satisfaction. For each of the seven questionnaire items included in Table 5.4, Ranger accommodation was rated more highly than non-Ranger which suggests that, overall, there is a significant gap between the housing package available to Ranger employees and that available to others. For a start, the weekly rent for a Ranger house ($15.50) was a fraction of the rent charged to government housing residents (median rent $44.00, with some paying more than twice that). Further, as we have noted, Ranger housing includes full airconditioning and an electricity subsidy covering running costs, while government houses rely on louvres and natural ventilation—unless residents wish to install their own *un*subsidised airconditioning.

The housing package for single mine employees (SPQ accommodation plus all meals at the mess for $30.00 per week) was also considerably less expensive than that available to most government SPQ residents whose median weekly rent of $35.00 did not include meals. Government SPQs also lacked built-in airconditioning and, as indicated above, were designed and sited in a way which created privacy problems for the occupants.

Contrasts between company and non-company housing have been noted in many other mining towns (for example, Brealey and Newton, 1978: 77; Robinson, 1984), and Jabiru is far from unique. The difference between housing packages is a sore point for some non-company households, however, and is undoubtedly the reason that housing satisfaction was relatively higher for those living in Ranger accommodation.

In the following section we turn to an evaluation of a specific aspect of the physical planning of Jabiru, the decision to disperse SPQs in residential neighbourhoods rather than concentrating them in their own area or 'ghetto' in the community.

Dispersing the SPQs

The most common way to accommodate single persons in mining towns is to confine them in an enclave separated from housing for married employees by a buffer-zone of playing fields, the town centre or a highway. In Utah coal communities in the Bowen Basin, singles have even been located close to the mines some 20 km or more out of town. The rationale underlying decisions to separate them in this way, whether within the community or completely outside it, is that aspects of the behaviour of singles may be offensive to married persons and, in the interest of community relations, singles should therefore be separated from families (Neil et al., 1982: 9; Oeser, 1981: 209). In addition, there is an assumption that many (if not all) singles would prefer to be off on their own in any case.

Planning for Jabiru consciously rejected this 'yobbo' theory of social management and proposed instead to include the single persons'

Single Persons Quarters (SPQs) *Ranger SPQ with six units per building* (above left)*; government SPQ with three units in this building* (below left)*; front view of government SPQ* (above).

accommodation within the community, dispersed in married neighbourhoods (see Map 14). To minimise the visual impact of SPQs, they were designed as one-storey buildings and generally were sited at the end of culs-de-sac. All Ranger SPQs have six units per building and, because they do not include cooking facilities, are located within a ten-minute walk of the Ranger mess via the town's pathway system. The location of government SPQs was not constrained by a need to be near the mess, so they are even more dispersed.

To gauge how well this arrangement has worked, we asked Jabiru residents the following question:

> Housing for single people in mining towns is often kept separate from housing for families. In Jabiru the single persons quarters are located in each of the neighbourhoods rather than being kept separate. Do you think the arrangement in Jabiru has worked out *very* well, fairly well, *not* very well, or not at all well?

Responses to that question are summarised in Table 5.5 and are tabulated by the respondent's proximity to SPQs. Residents of SPQs

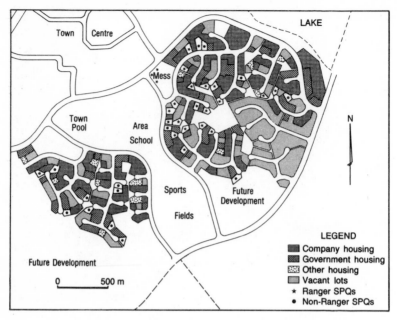

Map 14 *Distribution of single persons quarters, Jabiru.*

were the ones who were most convinced that locating them in residential neighbourhoods had worked out very well. In the community as a whole 92 per cent of respondents judged that dispersal of SPQs had worked at least 'fairly well', and almost half said 'very well'. It is interesting that residents who lived closest to SPQs (either immediately adjacent or within three lots) were more likely to say that the arrangement had worked out 'very well' than were those who lived further away. On the other hand, those living nearest to SPQs also included the people who were most likely to report that things had not worked out at all.

We did not ask respondents to differentiate between Ranger and government SPQs, but separate tabulations were run which classified residents on their proximity to Ranger SPQs and to government SPQs. While just over 80 per cent of those living next to or within three lots of a Ranger SPQ said that the arrangement had worked 'fairly' or 'very well', 93 per cent of those living that near to a government SPQ responded in a similar fashion.

The bottom part of Table 5.5 summarises reasons people gave when they were asked why the arrangement of SPQs was (or was not) working out. By far the most frequent response was that mixing single

Table 5.5 Reaction to SPQs dispersed in neighbourhoods
(percentages of respondents)

	Lives in SPQ	Next to SPQ	SPQ w/i 3 lots	SPQ on cul-de-sac	Not near SPQ	Total
How well has dispersal of SPQs worked?						
Very well	61	44	44	37	37	48
Fairly well	38	41	47	45	51	44
Not very well	1	8	5	18	8	6
Not at all well	—	7	4	—	4	2
Why dispersal has worked						
Positive reasons						
'Good idea' to mix single and married people	79	61	45	21	61	61
More interaction when mixed	22	22	15	41	24	22
Makes singles part of community	22	26	20	24	10	18
Singles behave better this way	5	13	10	18	6	11
Neutral response						
Haven't heard of any problems	13	17	60	65	39	25
Number of respondents[a]	70	36	33	28	80	247

Note: a Only respondents living in the Jabiru townsite are included in this table. This excludes those in Jabiru East, the caravan park and the Nationwide demountables who would have had less exposure to the dispersal of SPQs in the Jabiru townsite neighbourhoods.

person housing with married housing was basically a good idea. Other frequent responses were that there was enjoyable social interaction between singles and marrieds, that singles were made to feel more a part of the community, and that singles tended to behave better when they were living around families.

Negative reactions to the dispersal of singles around the community were too infrequent to warrant their inclusion in Table 5.5. Although women were slightly more likely than men to say that dispersing the SPQs had not worked 'at all' or 'not very well' (nine per cent versus six per cent), married men tended to be more outspoken than women in their comments on the siting of SPQs. This occurred in some cases because marriages had evidently come under threat from a resident of a nearby SPQ. An indication of some of the problems and the virtues of siting SPQs in residential neighbourhoods are included in the selections below.

Works 'very well'
Male, 34, single, lives in Ranger SPQ

> You have contact not only with the single people. You don't feel like a separate race.

Female, 34, single, lives in Ranger SPQ

> People are segregated in places like Newman [Pilbara iron ore town]. It's like a jail there.

Female, 37, separated, lives on cul-de-sac with Ranger SPQ

> It's made the community a much more friendly place. There's not the segregation here. They all seem to behave because they are mixed in.

Works 'fairly well'
Male, 34, married, lives in government house adjacent to government SPQ

> They have been no more disruptive than the couple on the other side. There is some difference because they change residents more often.

Works 'not very well'
Male, 50, married, lives in Ranger house adjacent to Ranger SPQ

> Wrong way in my opinion. It's hard for them to get to the mess in rainy weather. Mess and the SPQs should be together. But it's good...less fights and arguments.

Works 'not at all well'
Male, 34, married, lives in Ranger house adjacent to Ranger SPQ

> Too noisy. They've got no respect for people living next door. I know some groups that are good. This one next door seems to have all the drunks in it.

Male, 33, married, lives in Ranger house within three lots of government SPQ

> Just doesn't work. Two different lifestyles. Parties and everything else...always noisier parties. Increases problems in marriages when singles are mixed in with families. Increases socialising a lot more—which leads to the other problems.

In addition, several parents were concerned that locating SPQs at the end of culs-de-sac made the streets less safe for children because of the speed at which some singles drove their cars and motorcycles. Even so, most of these people also felt that dispersing the singles in Jabiru was, all things considered, a good idea. In another street families who endured an all-night party on the roof of a neighbouring Ranger SPQ—complete with furniture and stereo—were guarded in their enthusiasm for having these quarters nearby (in the case of that party, one respondent living in a Ranger house reported keeping the home airconditioning on to muffle the noise—an alternative not available to those in government houses).

To summarise, while over 90 per cent of residents said the decision to locate SPQs in Jabiru's neighbourhoods had worked out 'fairly well' or better, the selection of interview responses above indicates why the dispersal policy was not universally applauded. Some singles occasionally behaved in ways that their neighbours found particularly obnoxious. Some marriages were under stress and it may be that the proximity of SPQ residents contributed to the extent or the timing of those difficulties. On balance, however, the survey results show a level of acceptance from both singles and marrieds that provides strong support for the dispersal of SPQs in mining towns in the future.

In the following section we turn from housing to questions concerning the community as a whole and will return to Ranger/non-Ranger and married/single comparisons. We have already found non-Ranger households to be relatively dissatisfied with their housing and we wish to determine if that pattern of response is repeated in ratings of community facilities and services.

Resident ratings of Jabiru

One of the main attractions of building a new, independent, mining town instead of attaching growth to an existing community is that new towns do not possess an entrenched community structure of local people, facilities and services. There is an opportunity for planners and their clients to decide on a town's physical layout and the mix and location of town amenities, but the fact that facilities are *new* is no guarantee that residents will be satisfied with them. Whether they will be evaluated highly depends on a number of factors in addition to their newness, including not only the quality of the facility or service in question, but also the experiences and expectations that residents bring to a new community from their previous towns. People who moved to Jabiru from somewhere else in the Northern Territory, for example, tended to rate Jabiru more highly than people who arrived from other parts of Australia, particularly those coming from Queensland.

Residents were asked in the survey to use a five-point scale (from 'excellent' to 'poor') to rate eleven different facilities and services in Jabiru. Table 5.6 summarises these results by presenting the percentages of respondents who rated these facilities as 'good' or 'excellent'. Six of the items were rated that highly by at least half of those responding. At the top of the list was the Jabiru pre-school

Table 5.6 Ratings of community facilities and services
(percentages of respondents rating facility 'good' or 'excellent')[a]

	Ranger married	Non-Ranger married	Ranger single	Non-Ranger single	Total
Pre-school kindergarten	92	88	89	93	90
Upkeep of public areas	78	87	86	91	84
Garbage collection	85	82	57	71	76
Sports and recreation facilities	76	74	54	63	64
Churches and religious facilities	65	51	51	38	55
Schools	50	59	40	72	52
Local shopping centre	50	46	40	46	46
Health and medical services	38	38	54	58	46
Children's playgrounds	34	42	48	55	42
Social clubs	32	41	38	28	34
Restaurants	15	10	18	13	14

Note: a Not all residents provided ratings of each of the eleven facilities. New arrivals in town were often unable to give an assessment of all of the items, for example, and households without children were less likely to venture a rating of playgrounds and schools. The total numbers providing ratings for each facility were: pre-school kindergarten (162), public upkeep (320), garbage collection (304), recreation facilities (305), religious facilities (188), schools (203), shopping centre (320), medical services (297), playgrounds (237), social clubs (295) and restaurants (287).

kindergarten and day-care centre—90 per cent rated it 'good' or 'excellent'. Upkeep of the public areas and garbage collection (carried out by the Gagadju Association under contract) were also rated highly by at least three-quarters of the resident households. At the bottom of the ratings were restaurants (only 14 per cent rated them 'good' or better) and social clubs (34 per cent 'good' or better).

A review of the table as a whole reveals few major differences in ratings based either on marital status or on Ranger/non-Ranger employment. There are some contrasts worth noting, however. For example, married households were more satisfied with their garbage collection than were those living in SPQs, particularly those living in Ranger SPQs. Possibly more interesting is that singles, who were more likely to be involved in active sports like football and cricket than the married households, were more critical of the town's sports and recreation facilities (responses to items elsewhere in the questionnaire indicated concern about the hard-baked surface of both the cricket oval and football field and a desire on the part of some residents for an indoor gym and squash courts). On the other hand, singles gave higher ratings than marrieds to health and medical services and to children's playgrounds. In both of these cases the higher ratings from singles may reflect a lack of knowledge on their part of the needs of families with young children in Jabiru.

Earlier in this chapter we found that households living in non-Ranger housing were less satisfied with their dwellings than those living in Ranger homes. No similar pattern of response occurred for the facilities and services listed in Table 5.6. In fact, in two instances—churches and religious facilities and schools—the non-Ranger respondents reported notably higher ratings than the Ranger residents. Additional tabulations (not detailed here) showed that the sex of the respondent was not significantly related to ratings of any of the facilities noted in Table 5.6. On the other hand, length of residence was a factor in several cases: newcomers gave more positive ratings to community upkeep, children's playgrounds and restaurants, while those who had lived in town for three or more years rated religious facilities more highly.

We did not have time in the interview to probe the reasons for all of the ratings of facilities that have been summarised in Table 5.6, but we did ask those who rated a facility 'poor' why they did so. Most facilities were rated 'poor' very infrequently; only health care (11 per cent), social clubs (11 per cent), children's playgrounds (15 per cent), and restaurants (46 per cent) were given ratings that low from more than seven per cent of the households. These complaints are not unique to Jabiru, of course, and data for Hedland summarised by Brealey and Newton (1978: 50-5) indicate that health care, playgrounds and restaurants are among the problems faced by residents in the Pilbara as well.

In Jabiru dissatisfaction with health and medical services was partly a matter of unfulfilled expectations. A hospital facility that was built in the town to handle the combined populations of three mines was downgraded to a clinic when only Ranger gained permission to commence operations. The resulting lack of a full range of medical services has created considerable inconvenience for some people as the following comments show.

Male, 36, married, two children

Bloody shithouse. Can't get medical help when you need it. Not up to standard. For expert opinion you have to go to Darwin. My daughter broke her leg on a Friday...would have had to wait 'til Tuesday for an x-ray.

Female, 31, married, two children

Only one doctor. Can't get in when you want to get in. Husbands who finish work at 4 have to take a day off [to see the doctor]. Pregnant mothers have the biggest problem...have to go into Darwin to have children. No overnight accommodation at the clinic. Prescriptions may not be available at the chemist. It's pathetic.

There were two social clubs in Jabiru at the time of our interviewing: the Jabiru Sports and Social Club (JSSC) located at the

Jabiru town facilities *Child care centre in town centre* (above left); *olympic-size swimming pool with picnic area* (below left); *Jabiru Cafe in town centre* (above).

west end of the town lake; and the Golf Club located on the south edge of the town (see Map 13). Although the latter has the demountable look of a construction camp wet mess, its clientele comes largely from managerial and salaried residents rather than the wages end of the continuum. The JSSC overlooks the lake in a substantial building and caters more for the needs of single wages employees. It has one of the two restaurants in Jabiru, is the place where Aborigines from the Manaburduma Camp come to buy alcohol, and was the venue for one or two noteworthy Friday night punch-ups in the first weeks of our interviewing period. Reasons for rating these social clubs 'poor' are typified by the following:

Male, 36, married, two children

It [the JSSC] is just a pub—a rough one. No good chef or management. You wouldn't take your wife to it for tea.

Male, 24, married, no children

I've been involved in other clubs that do things for the community. Here they're mainly into drinking beer,

The main reason for a 'poor' rating of children's playgrounds was that 'there aren't any' (or that 'there aren't enough'). By mid-1984, Jabiru had one adventure playground on the south side of the town lake and one being built in the centre of the open space in Jabiru's eastern neighbourhood. Much the same sort of comment ('there aren't any') resulted when we probed the very large number of 'poor' ratings of restaurants. At the time of the interviewing only two restaurants existed in town, but they clearly did not meet the needs of many of the residents. The one in the JSSC provided a limited range of *à la carte* dishes which, from our own experience, would be unlikely to attract many singles away from the mess if it were not for the availability of alcohol at the JSSC. The other restaurant was the town centre café. It opened only during shopping hours except for a 'pizza night' once a week. One respondent was loath to place all the blame on the restaurants, however.

Male, 40, married, no children

> [The food] varies from inedible to poor. It's the inhabitants' collective fault. The people here are notoriously stingy. People won't pay high prices for good food.

In addition to the ratings included in Table 5.6, the residents were also given an opportunity to rate the community as a whole. The question asked was:

> How do you feel now about this community as a place to live? From your own personal point of view would you rate Jabiru as an excellent, good, average, below average or poor place to live?

Responses to this question were probed twice ('Why is that?'; then, 'Anything else?') to flesh-out answers to the closed-ended ratings. Results in the first panel of Table 5.7 indicate that 69 per cent of respondents rated Jabiru 'good' (53 per cent) or 'excellent' (16 per cent). If one considers only 'excellent' responses, satisfaction is slightly higher among the married Ranger households and roughly similar across the other three groups. If 'good' responses are combined with 'excellent', the differences disappear. (Additional tabulations indicate that newcomers in town—1984 arrivals—were slightly less likely to rate the community 'good' or 'excellent' than those who arrived in 1981 or earlier: 63 per cent versus 72 per cent.)

Virtually the same five-point community satisfaction question has been used in a number of residential environments in Australia and overseas. Responses in Jabiru compare favourably with the findings from many of these studies. For example, a 1984 study of medium and higher density housing in suburban Sydney found from 59 per cent (Willoughby) on up to 83 per cent (Centennial Park) and 97 per cent (Wollstonecraft) rating their communities as 'good' or 'excellent'

Table 5.7 Satisfaction with the community
(percentages of respondents)

	Ranger married	Non-Ranger married	Ranger single	Non-Ranger single	Total
Community rating					
Excellent	22	14	13	11	16
Good	47	56	52	60	53
Average	24	26	25	17	23
Below average; poor	7	4	10	12	8
Reasons for ratings were					
Positive only	66	56	63	63	62
Positive and negative	12	17	16	20	16
Negative only	22	27	21	17	22
Positive reasons for ratings					
Attractive environment; Kakadu National Park	50	42	40	24	40
Lifestyle; small town atmosphere; away from the city	29	20	29	8	22
Friendly community; community spirit	18	18	19	32	21
Community facilities, services; planning	13	18	19	20	17
Job availability; personal economic reasons	17	9	7	18	13
General response n.e.c.; 'I just like it here'	22	20	14	28	21
Negative reasons for ratings					
Community facilities, services lacking	9	16	8	4	9
Dislike environment; remoteness; small town	6	10	10	1	6
Unfriendly community; lacks community spirit	9	3	3	2	5
Divisions in community	3	4	2	—	2
General response n.e.c. 'I don't like it here'	10	8	10	18	11

(unpublished data, School of Town Planning, University of NSW). A 1975 study which included residents in Woden, ACT, and in the Sydney suburb of Sherwood Hills, NSW, found that 87 per cent and 76 per cent, respectively, of those respondents rated their communities that positively (unpublished data, School of Town Planning, University of NSW). In the United States a study of 15, mainly suburban, new communities included one remote desert community, Lake Havasu City, Arizona, where 79 per cent of the residents rated their town as 'excellent' or 'good' (Zehner, 1977:43). In contrast, an earlier American study of four working-class communities in Maryland found only 29 to 54 per cent that satisfied (Zehner and Chapin, 1974:115).

The next sections of Table 5.7 focus on reasons given by residents for their ratings of the community. Multiple responses (both positive and negative) were coded from each interview so figures should not be expected to add to 100 per cent. In view of the generally positive evaluations of Jabiru on the excellent-to-poor scale, it is not a surprise to find responses to a probe of those evaluations overwhelmingly positive. In fact, 62 per cent of people we interviewed had only positive things to say about Jabiru, 16 per cent reported both positive *and* negative things, and 22 per cent made only negative comments about the community.

Table 5.7 should not be taken as an indication that the community is relatively free of problems, however, and we report in chapter 6 on an open-ended question specifically asking residents to identify 'problems and issues' in Jabiru. Having said that, the results in Table 5.7 do indicate that most residents were able to find several good things to say about living in Jabiru. Certainly the most frequently mentioned reason for a positive rating of the community was the attraction of the natural environment and the proximity of Kakadu National Park. Closely related to those responses were comments about the relaxed lifestyle possible in a small town well away from city life. Over 20 per cent of the respondents also mentioned the friendliness and community spirit they found in Jabiru, and somewhat smaller percentages mentioned the quality of one or more of the facilities and services available in the community.

Negative reasons elicited by the ratings probe also covered a wide range of issues, the most frequent being dissatisfaction with one or more kinds of facility in Jabiru. Six per cent of the respondents mentioned a *dis*like of the environment and/or Jabiru's remoteness and small-town atmosphere. In similar fashion, some people reported that they found Jabiru an *un*friendly place lacking in community spirit. Only two per cent of the respondents mentioned splits in the community (for example, Ranger versus non-Ranger; staff versus management) in answer to the community rating probe. We expected more responses in this category because of the strike only months before which, if it accomplished little else, certainly highlighted differences between groups in Jabiru.

Several responses have been selected from the interviews to provide a flavour of the range of factors that were of importance to respondents rating Jabiru.

'Excellent'
Female, 56, married, no children

Peace and quiet. A secure job. I just like it. The facilities are very good for a mining town. Darwin is there if you need it.

Male, 30, single

As a single bloke I'm accommodated well and fed well. The local environment is a national park! No smog, no hustle and bustle.

'Good'
Male, 34, single

This is a friendly place. People say 'G'day' when you don't know them from a bar of soap.

Female, 42, married, two children

Services are not as good as in a city, but as far as enjoying life, it's as good as you get anywhere...the climate...being in the middle of the Kakadu National Park.

'Average'
Male, 21, single

Nothing for a single guy except to go to the club and piss your money against the wall. Not enough women and it's a long way to the next town.

Female, 19, de facto

It doesn't cater for women. It only caters for the workers at the mine.

'Below average'
Male, 54, single

It's not a complete town. We don't even have a cemetery. No retired people. Teenagers are off at school [out of town]. People here are transients. There's no real continuity.

Female, 32, single

Lack of community spirit. It's hard to define it. I don't see people getting involved. They seem to believe they should have everything given to them without making an effort.

'Poor'
Male, 35, divorced

The town is divided...staff and award don't mix. Even staff kids aren't allowed to play with award kids. Nothing for people to do except go to the club...which seems to lead to marriage breakdowns.

The section has focused on a series of ratings of the community and the reasons behind these ratings. We noted a few instances where marital status, or Ranger/non-Ranger, appears to make a difference in the evaluations, but in general there are many more similarities in ratings across marital status and employer groups than there are differences. The relative absence of such differences is in contrast with the earlier section on housing in which those variables did prove to be important.

At the beginning of this chapter we noted our concern, prior to undertaking the survey, that recent industrial problems in Jabiru might seriously complicate interviewing. Labour management relations have been an important dynamic in some Australian mining towns (for example, Moranbah, Qld in Williams, 1981; and Whyalla, SA in Kriegler, 1980), and we had expected comments about detrimental effects on the community from the FMWU and ASE confrontation with ERA. Surprisingly, the topic of community divisiveness and union versus management (or even Ranger versus non-Ranger) barely rated a mention in the generally positive responses about the friendliness of Jabiru and attractions of Kakadu and the surrounding natural environment. The survey did provide other opportunities to comment on the community, and in chapter 6 we will find that tensions between different sections of the community were raised as an issue by a small but significant number of respondents.

Factions and divisiveness in Jabiru were not dominant factors in the evidence contained in the present chapter, however, and a more fitting conclusion would be the response one resident gave for rating Jabiru a better community than the last place he lived.

> We like it here. The weather...the people...[and] it's an easy place to bring up kids.

6 Problems and prospects: the community response

> People should take more interest in the place. The majority don't care.
> They're only here for their quid and that's it.
> *(Male, 32, married,* Jabiru Community Survey)

Identifying issues of importance to residents of a mining community can be done in several ways. Sufficient material has been published on Australian resource-based towns to provide a sizeable list of issues worth investigating (Stockbridge et al., 1976; Australian UNESCO Seminar 1976; Brealey and Newton 1978; Neil et al. 1982) and there was no reason to expect that a survey in Jabiru would uncover problems that had not been experienced elsewhere, with the exception of its location in the Kakadu National Park and its experiment with the Jabiru Town Development Authority (JTDA). Another source of information was our regular visits to Jabiru to talk to people with differing personal and organisational perspectives on the community.

Many of those interviewed in 1982-84 were involved in town, mine or National Parks administration and found the relationship between town and park and the role of the Development Authority engaging topics. Other things which arose in discussion included the possibility of more tourist development, concern for high school education, teenager activities, marriage problems, and the circumstances of Aborigines in the Manaburduma campsite.

We could not tell if the majority of Jabiru residents shared these interests and concerns, however, and the community survey was therefore designed to give residents several opportunities to tell us what *they* thought the main issues facing the town were. This chapter looks at these community issues—both those which were addressed by specific items in the questionnaire and those which arose in response to open-ended probes of the residents' concerns.

Problems and issues

Not all topics can be covered first in a questionnaire. The interviews in our survey began by asking why people chose to move to the

135

community and how they rated the town and its facilities. An open-ended question then followed: 'In your opinion, what are the most important problems and issues facing Jabiru at the present time?' In view of this it was not surprising to find that those who were critical of the town's facilities often reiterated those points when asked the question about the community's 'problems and issues'. Had the interviews *started* with an open-ended probe into 'problems and issues' in Jabiru a somewhat different mix of responses might have resulted with fewer facility-specific mentions.

In the event, the sequence of questions in the interview clearly did not limit respondents to comments about facilities and services in the town. The results of the query about 'problems and issues' are summarised in Table 6.1 and it is interesting to note that one in eight interviews mentioned four (or more) issues. Percentages do not sum to 100 because up to four responses were coded per interview.

The residents' agenda of community problems and issues covered a wide spectrum, though only those subjects which were mentioned by at least five per cent of respondents are included in the table. Many of the items thereby omitted addressed one or more social aspects of the community and ranged from concern with marital problems and the status of Aborigines to the shortage of single women in town.

Problems with shopping were mentioned by 18 per cent with the complaints including things like high prices, the lack of choice and the monopoly position of most Jabiru businesses. An independent confirmation of the high cost of groceries is provided by *Choice*'s (June 1985) '6th Annual Supermarket Survey' which put Jabiru prices higher than five of the six Darwin stores surveyed and, on average, the most expensive of the 20 cities included in the review. Married households were much more likely than singles to mention shopping as an issue, and non-Ranger singles (who have to prepare their own meals) were predictably over twice as likely as their Ranger counterparts (who eat in the mess) to mention it as a problem.

The next topic in order of concern was government policy on uranium, raised by 15 per cent of the respondents, which reflected discussion in the country at that time, particularly within the ruling Labor Party in Canberra. The possibility that the Ranger operation might be closed down, however unlikely, served to reinforce awareness in the town of its dependence on outside forces.

As we have noted, we expected a number of respondents to mention divisions between wages and staff, and Ranger and non-Ranger, households in Jabiru because of the recent strike. Only 14 per cent did so, and although some of them were especially upset at the problems they saw in the community, this is a low proportion compared to other studies. For example, Stockbridge et al. (1976: 20) found that almost

Table 6.1 Community problems and issues
(percentages of respondents)

	Ranger married	Non-Ranger married	Ranger single	Non-Ranger single	Total
'Most important problems and issues facing Jabiru'					
Social					
Community divisions[a]	12	16	14	16	14
Activities for teenagers	14	11	5	18	12
Lack of community spirit	8	8	6	5	7
Drinking; alcohol	13	6	4	—	7
Economic					
Future of uranium mining; government policies regarding uranium	19	13	10	17	15
Facilities					
Shopping/commercial[b]	23	23	7	16	18
Health/medical care	16	18	6	—	10
Tourist facilities[c]	9	17	7	10	10
Recreation	7	8	13	6	8
Schools	10	10	3	—	6
Other mentions					
Town administration	10	10	10	9	10
Dogs running loose	5	11	3	5	6
Remoteness	4	8	13	4	6
Housing quality	6	3	4	8	6
Number of problems and issues mentioned					
None	4	7	7	1	5
One	28	26	27	32	28
Two	31	29	42	24	31
Three	21	25	16	29	23
Four or more	16	13	8	14	13

Notes: a Community divisions include those based on Ranger vs. non-Ranger, staff vs. wages and unions vs. company.
b Includes comments about high prices, lack of competition and choice among stores, lack of variety in stock and needs for specific types of stores.
c Includes general mentions of the need for tourist facilities and specific mentions of caravan park, camping ground and motel accommodation.

half of their respondents mentioned social stratification as a problem in the Roebourne Shire of the Pilbara, and Williams (1981) demonstrated that class and stratification issues pervaded working and community relationships in her study of Moranbah.

We noted in chapter 5 that our questionnaire was designed to avoid becoming an industrial relations issue in Jabiru, and there is no doubt that questions directed at the strike and its effects would have increased the proportion of those who felt that divisions in the community were an important issue. Interpreting the significance of 'only' 14 per cent mentioning community divisions is rather like trying to decide if a glass of water is half-full or half-empty. It is sufficient to say that although class and other divisions do not appear as central an issue in Jabiru as in other mining communities, the issue of splits among groups in the community outpolled all but two other issues. Further, if it is assumed that some or all of those who mentioned lack of 'community spirit' as an issue also referred to divisiveness in the community, then community divisions would have been the most frequent response.

Of the remaining issues listed in the table only four were mentioned by at least 10 per cent of the residents: more activities for teenagers, problems with health and medical care, the need for added tourist facilities and problems with town administration. More similarities were found across the four categories of respondents in Table 6.1 than there were differences. Three contrasts should be noted. As might be expected, concern with schools was an issue almost solely among married respondents and, as suggested in chapter 5, comments on medical care also came predominantly from that group. Secondly, singles were less likely to report drinking or alcohol an important problem. And finally, non-Ranger respondents commented most often on the need for tourist facilities, undoubtedly a reflection of the fact that they would be the ones most likely to capitalise on any significant increase in developments in that sector.

Eleven selections from the interviews follow which provide not only an indication of the variety of reactions to the question about 'problems and issues', but also an insight into the substance of the percentages in Table 6.1.

Male, 54, single, non-Ranger professional

Boredom. The grog. They don't have other people to care for. It's an unreal situation. There's no awareness of the needs of other people. (Anything else?) This is a very class conscious, cliquey place.

Male, 34, married, Ranger wages employee

A line down the middle of town between staff and wages. It shows in the Sports and Social Club and the Golf Club. Extra money for freight is a lot of bull. It's cheaper to get things from Perth than from Darwin. There's no competition, just monopolies.

Female, 38, married, husband Ranger wages employee

The company and its disregard for the welfare of its employees. The workforce are treated as automatons. No taxis or buses. The dissatisfaction of the teaching staff because the government housing isn't up to company standards. High rents.

Male, 33, separated, Ranger wages employee

The cost of living. Things are more expensive than in Darwin. Everything is a monopoly here. If you want something for your car you have to go to Darwin or go to them here and pay exhorbitant prices and get lousy service.

Male, 36, single, Telecom technician

JTDA is not all that flash. The attitude of most people has to change. The town is too Ranger-orientated. There are three cliques—government workers and Ranger staff and wage workers. Family relationships are pretty unstable.

Female, 47, married, non-Ranger professional

Not enough offerings for teenagers. The cinema closed down. That had its problems, but at least it gave them a meeting place that was public. Without it there's X or R rated movies on the video or worse.

Female, 31, married, husband Ranger wages employee

Isolation. A lot of marital problems. Do I send my daughter away to high school? Opinion is that the secondary school here is not satisfactory. Shopping is always a problem. Could do with a family counsellor here, though many bring their problems with them.

Male, 45, married, non-Ranger executive officer

Most of the community are young people away from home for the first time. They have no family to fall back on. Also... lots of money spent too easily. (Anything else?) Lots of family breakups. Too much leisure... leads to drink which leads to social problems. Everything is done for everybody. It's all provided. I'm a member of a service club and it's damn hard to find any worthwhile projects to do.

Female, 28, married, husband OSS professional

Dogs. Too many dogs. They're left to roam the streets. Never had so much trouble with dogs as I have had here. You can't have cats so everyone has two or three or four dogs.

Male, 22, single, OSS technical assistant

I think the place is still very young and needs more time to develop more character. There are so few problems because we are very sheltered from outside. There's not any unemployment and so on. People complain of isolation, but maybe it's a blessing in some ways.

Female, 33, single, Ranger staff employee

It is a mining town. People arrive with views to leaving. You just have to accept that.

The sheer variety of subjects mentioned in Table 6.1 and in the detailed responses in the interviews is itself instructive. It means, on the one hand, that no single aspect of Jabiru has fallen so short of expectations that it elicits reflex complaints about that issue. On the other hand, most respondents were able to come up with at least two 'problems and issues', and while this may only be an indication of competent interviewing, it does suggest that the new Jabiru Town Council will have a variety of things to deal with during its first years in office.

At the beginning of this section we noted that the relationship of town and National Park and the use of a development authority for town administration were topics of particular interest to community leaders contacted during earlier visits to Jabiru. The survey questionnaire addressed both of these things and we will shortly consider them in more detail, but it should be recognised that relatively few residents considered those topics as 'important problems and issues facing Jabiru at the present time'. Only 10 per cent commented on the Development Authority or the town administrators, 10 per cent saw a need to expand tourist facilities, and less than three per cent mentioned the limits on use of the Park or controls on fauna and flora in the townsite imposed by Park regulations.

Kakadu National Park

There is a sign on the Arnhem Highway driving east from Darwin announcing entry to the Kakadu National Park which is virtually within sight of the highway marker indicating that there are still 100 km to Jabiru. South and east of town the road to Pine Creek makes its way over Nourlangie and Jim Jim creeks and covers over 80 km from Jabiru before reaching the Park boundary. When present work on that road is complete Jabiru residents will have a second all-weather surface route out of the Park. The eastern edge of the Park is much closer to the town, of course, but the East Alligator River crossing, some 30 km northeast of Jabiru, also forms the boundary with Arnhem Land which is prohibited territory for those without a permit. The latter are given infrequently (Palmer, 1985).

Most Australians have to drive at least an hour to get *into* a state or national park. Jabiru residents now need to drive that far to get *out of* the Kakadu National Park. Prior to the proclamation of Stage Two of the Park (see Map 1) the area immediately north of Jabiru fell outside the Park and its westernmost edge was only 40 km away at the South Alligator River. For the many Jabiru residents who enjoyed the freedom of hunting, fishing and camping where they wished in those areas, enlarging Kakadu was hardly an occasion for celebration.

Sportsmen are not the only ones inconvenienced by the enveloping presence of the Park, and it may be assumed that few, if any, residents have lived in places where 'general tourist accommodation' is prohibited; where 'no animal may be taken into the town' other than dogs; where 'No plants shall be brought into the town without the consent of the Director' [of ANP&WS]; and where 'Regular inspections of the town and its operations will be undertaken to ensure compliance with regulations and conditions' (Australian National Parks and Wildlife Service, 1980: 301-304). Given these conditions it is difficult to interpret the statement which follows in the Plan of Management that 'An understanding of. . . national park policies can make the opportunity to live in Kakadu a fascinating privilege' (Australian National Parks and Wildlife Service, 1980: 305-306).

Despite constraints on people's behaviour in the Park *and* in the town, it will be recalled from chapter 5 that the attractiveness of Kakadu and the environment around Jabiru was one of the main sources of resident satisfaction. We were aware that living within an enclave of the Park seemed to be a mixed blessing, and a key question in the survey asked respondents to weigh the pluses and minuses:

> Some people feel that having the Kakadu National Park all around Jabiru is an advantage while others think that it is more of a disadvantage. How do you feel about it? Would you say that it is more of an advantage than a disadvantage or more of a disadvantage than an advantage?

The responses appear in Table 6.2 together with replies to specific questions about the Park's advantages and disadvantages. Two of the sections in the table are of particular interest. First, close to 70 per cent felt the Park was more of an advantage than a disadvantage, but second, almost 80 per cent had both bad and good things to say about the Park. In fact, only 13 per cent indicated that there were *no* disadvantages. The plus factors of the Park are easy access to recreation and, simply, proximity to a beautiful part of Australia. Twenty-six per cent also mentioned long-term benefits for the environment due to National Parks and Wildlife management. In addition, nine per cent of the residents thought that attracting tourists to the area benefited the town, some because it increased the variety of people in and around Jabiru, and others because it increased the amount of money spent in the community by outsiders.

Interestingly, 16 per cent of respondents also said that attracting tourists to the area was one of the main *dis*advantages of Kakadu. As several quotes included below indicate, some felt that there were already so many tourists around that it was hard for 'locals' to find a quiet place for themselves in the Park, particularly in the 'dry'. By far

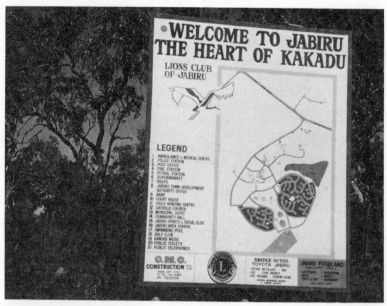

Kakadu National Part *Town sign outside Jabiru* (above left); *Jim Jim Falls in the Dry* (below left); *four-wheel drive track near Jim Jim Falls* (above).

the greatest grounds for complaint, however, were the limitations on where people could go in Kakadu and what they could do. Although these were clearly the main disadvantages as seen by each of the four resident groupings, single respondents appear to resent limitations more than the married households. A small proportion (three per cent) of the 50 per cent who were concerned about their lack of freedom in the Park were pointedly upset with the fact that an increasing number of areas were being closed off to whites because of the needs of the traditional Aboriginal owners of the land. One person estimated that access was already restricted to half or more of the National Park and that that figure was continually on the increase. The first responses below are indicative of this reaction.

Advantages and disadvantages of Kakadu

Male, 34, married

> Too many restricted areas. Too many restrictions. The Aboriginal community can go anywhere. They can come and live in the mining town...we can't go into their territory.

Table 6.2 Kakadu National Park
(percentages of respondents)

	Ranger married	Non-Ranger married	Ranger single	Non-Ranger single	Total
The park is					
More of an advantage	61	76	68	73	68
More of a disadvantage	35	18	24	24	27
Neither; both	4	6	8	4	5
Main advantages of the park					
Recreation	38	44	38	26	36
Beautiful area	32	37	38	34	35
Park management	24	27	17	40	26
Attracts tourists	14	3	6	9	9
Main disadvantages of the park					
Limits on park use for town residents; no limits for Aborigines	46	43	57	58	50
Attracts tourists	17	13	16	16	16
Restrictions on pets and plants in town	19	17	7	11	14
Summary of responses					
Only advantages mentioned	14	18	16	7	13
Both advantages and disadvantages mentioned	78	72	80	81	78
Only disadvantages mentioned	8	10	5	12	9
Frequency of park use					
Once or twice per week	20	22	16	21	20
A few times per month	34	39	23	31	32
Once per month	18	13	24	13	17
Less often	26	18	24	18	22
Never	3	8	13	17	9
Main activity on last visit					
Fishing	34	29	33	25	30
Sightseeing	21	18	15	19	19
Picnic; barbeque	15	16	9	6	12
Camping	10	11	14	12	11
Relaxing	3	4	7	11	6
Bushwalking	3	3	5	5	4
Other	15	19	16	22	18

Male, 36, married

Limits recreational activities in terms of hunting and shooting and where you can go. It's really managed as an Aboriginal reserve rather than a park...and managed from Canberra. All the things I don't agree with.

Male, 32, married

It's not the same as it used to be. You used to be able to go everywhere. And there are tourists everyplace. Maybe we were too spoiled at the start.

Male, 57, married

If National Parks weren't here these blokes with their 4-wheel drives would tear this place to bits. There would be guys out shooting their guns all night.

Female, 35, single

Too many [ANPWS] rangers. Too many rules. Can't do much poaching.

Female, 29, married

You can't buy booze around here. There's no hotel, No petrol for tourists after hours. Tourists stop at our house to ask where to go!

Female, 33, single

All the scenic spots have controls over them. What are left have tourists all over them.

Male, 35, single

A million and one tourists in the dry season. You feel like an animal in a zoo.

Whatever reservations Jabiru residents held about the existence of the Park at their doorstep, the final panels in Table 6.2 show that they use the Park often. Over half (69 per cent) said that they visited the Park at least once a month with 20 per cent going once or twice a week on average. Visitor usage during the dry season (which included our period of interviewing) is higher, however, and when a question was asked about 'the last time you visited Kakadu', 31 per cent were found to have gone on an outing within the last week. Most of these excursions into the Park involved more than one activity—fishing and relaxing or camping and bushwalking, for example. The final entries in the table indicate that fishing was the primary purpose of 30 per cent of these recreational trips followed by sightseeing, picnicking and camping. An 'other' category (18 per cent of the responses) included activities like boating, swimming, visiting Park Headquarters and bird photography.

To recapitulate, the Kakadu National Park has several important roles to play for the residents of Jabiru. Its beauty and the recreation opportunities it provides are appreciated and used frequently by many of the town's households, and just over two-thirds of the survey respondents saw its proximity to Jabiru as more of an advantage than a disadvantage. Most of the people interviewed commented on the disadvantages of Kakadu and its regulations as well—cat and horse lovers missed having cats and horses, gun lovers missed shooting, campers and fishermen were frustrated at not being able to camp and fish in areas closed off to serve the needs of traditional Aboriginal owners—and several people mentioned problems facing tourists in Jabiru if they wanted a place to stay the night (not available), to buy petrol or food after hours (not available) or to buy alcoholic beverages (not available to people who are not *members* of the Golf Club or Jabiru Sports and Social Club).

Although one disenchanted respondent summed up the feelings of many residents when he said that the Park would be better named 'Kakadon't' than Kakadu, it is important to remember that 91 per cent of the respondents saw at least some advantages to having Kakadu nearby, that 68 per cent said the Park was more of an advantage than a disadvantage and that, for many of these people, the proximity of Kakadu was one of the most attractive aspects of living in Jabiru.

The Aboriginal community

The way in which Aborigines are perceived in and around Jabiru is complex. It primarily revolves around two factors related to their status as traditional owners in the Kakadu area. First, as noted in the previous section, attempts have been made with assistance from National Parks staff to specify areas of the Park as off-limits to whites because of the area's importance to local Aboriginal people. As a result, a number of billabongs and camping areas popular with Jabiru residents have been declared out-of-bounds (see photograph p 73) and this has caused considerable resentment among some townsfolk. The second factor causing comment in Jabiru is the payment of royalties derived from uranium mining to eligible traditional owners. Some residents question the fairness of Aborigines receiving such 'unearned' income. Closely related to this is a concern about how and where the money is spent and, more specifically, the role of alcohol in the Aboriginal community.

Respondents were asked whether they considered too much, too little or the right amount had been done for Jabiru's Aborigines. They were then asked *why* they thought that 'too little' or 'too much' had been done. Responses are summarised in Table 6.3 and show a clear difference of opinion on the first question. Nineteen per cent said that 'too little' had been done; 24 per cent said that 'too much' had been done. Sixteen per cent of the residents replied with a 'don't know' and the remaining 41 per cent chose the middle ground of 'about right'. During the interviewing the 'about right' choice frequently arose after

Table 6.3 Aborigines in Jabiru
(percentages of respondents)

	Ranger married	Non-Ranger married	Ranger single	Non-Ranger single	Total
What has been done for Jabiru's Aboriginal community has been					
Too little	15	23	16	25	19
About right	44	39	40	38	41
Too much	20	24	24	30	24
Don't know[a]	21	15	21	8	16
'Too little' in what way? (percentage of those who said 'too little')					
Better housing	24	23	40	11	23
Training needed to handle finances	46	14	10	16	23
Need to change attitudes of whites	—	17	20	32	18
Better education	6	18	20	19	16
Better health care	12	12	—	22	13
Need employment	17	23	—	—	10
Need help with alcohol	12	9	20	3	10
'Too much' in what way? (percentage of those who said 'too much')					
Handouts discourage developing sense of responsibility	43	55	35	36	42
Too many are telling Aborigines what to do	14	21	39	9	20
Handouts to Aborigines are unfair to whites who work	28	17	14	6	17
Aborigines are treated too leniently by law and by Ranger	14	4	7	30	15
Aborigines damage things because they don't appreciate their value	19	9	—	15	12

Note: a 'Don't know' was not included as an option in the question read to respondents, but was available for interviewers to tick on the questionnaire.

a pause and a comment such as 'I don't really know what should be done...but I guess what has been done is "about right."'

The probe of why respondents thought that 'too little' had been done identified a number of concerns, many of which underlined difficulties the Aboriginal community has dealing with aspects of (and expectations from) white 'European' society. Reasons mentioned by at least 10 per cent of those responding are included in the last two sections of the table. Percentages add to more than 100 because up to three reasons were coded per respondent. To reduce the Aborigines' problems residents suggested a need for special training to enable them to cope with both family and business finances, for better education, for more jobs and for help with problems associated with alcohol. Better housing and better health care were also mentioned as areas where improvement was needed. Eighteen per cent of those who said 'too little' had been done focused on attitudes and expectations held by whites and said that these needed to be less censorious and more understanding in the future. Single respondents were more likely than marrieds to see a need for changing white attitudes; marrieds were more likely to suggest more employment for Aborigines was needed.

When residents were asked why they felt 'too much' had been done, by far the most prevalent reason was that so many people and organisations were involved with Aborigines that they were being pressured into a passive role instead of taking active responsibility for their own welfare. One-fifth of these respondents simply said that too many people were telling the Aborigines what to do.

Like those who said 'too little' had been done for Jabiru's Aborigines, respondents who chose the 'too much' alternative and then mentioned the need for Aborigines to develop greater responsibility were largely sympathetic to the Aboriginal population and the problems they faced. Not all respondents were sympathetic, however, as the remaining categories in the table show. Some, for example, were incensed about 'unearned' royalty income; others were upset because they thought both Ranger and local police tended to allow Aborigines to get away with behaviour such as absenteeism and public inebriation which would land white residents in serious trouble. Many of these people would probably say that they had nothing against Aborigines 'as long as they were treated the same as whites'. We also encountered a small number of respondents, about three per cent of the total sample, who saw no reason for Aborigines and whites to be treated equally and felt little or nothing should be done for the Aboriginal population.

These findings parallel those of Stockbridge et al. (1976: chapter 7) whose data indicate that people in the Pilbara felt that the main areas in which something should be done for Aborigines were education 'to aid in assimilation', providing jobs, and housing. Just under 20 per cent of the Roebourne Shire respondents said they 'didn't know' what should be done (also very similar to the Jabiru responses). More than 10 per cent of those in the Pilbara said 'nothing' should be done.

A selection of responses arising from probes in the Jabiru survey of the 'too little' and 'too much' answers appears below.

'Too little'
Male, 31, single

> More responsible involvement in the town...like to see them elected to the town council. There is a very real separation between the Aboriginal and white community—like two cells. I'd like to see more involvement of the groups together.

Female, 17, single

> Better housing...their own design. Their kids should be encouraged in school. Should have the same advantages as the white people in the club [JSSC]. Aboriginal women can't go in without being signed in. Men go in all the time.

Female, 32, single

> [They need] a reasonable health service so they are not starving every week and their children are not having health problems. They need an improvement in self image. The public health facilities are shocking and health education is poor. Access to alcohol is too easy.

'Too much'
Male, 35, divorced

> Too much in the wrong direction...should be put into self-designed houses. Children should be encouraged to attend school.

Female, 45, married

> Free education. Big mobs of money. They get so much money per fortnight if their children go to school. Our kids don't get it. They scream about the housing, but I've seen what they do to the housing.

Male, 56, married

> They're getting more than anyone. They should get off their backsides. If they want land, give it to them but don't do anything for them. They can work for a living...and they should.

Inequity is frequently in the eye of the beholder and it is apparent from Table 6.3 and the verbatim responses above that there is a considerable difference of opinion in Jabiru about the circumstances of the Aboriginal community and how they should be treated. Our results suggest, however, that if greater numbers of Aborigines were gainfully employed in Jabiru there might be less resentment toward them. At the time of the survey there were several sources of employment for Aborigines in the immediate area including the Gagadju Association, National Parks, the Mudginberri abattoir,

Ranger and a few government departments and agencies in town, but the total number of positions employing Aborigines would, as far as we could tell, have been only 50 or less. Significantly expanding the number of job openings beyond that level would require another employment base—another mine, for example, or increased commitment to tourism—but even then the problem of matching skills of unemployed Aborigines with job openings remains, and it is unclear how many unemployed local Aborigines could be enticed into the workforce in any event. It is instructive to note that a recent 'Study of the Impact of Tourism on Aborigines in the Kakadu Region' prepared for the Northern Land Council (Palmer 1985) dwells on possible detrimental effects of increased tourism and does not address the issue of Aboriginal employment. We may infer, therefore, that Aboriginal unemployment will continue to be a major issue among Jabiru residents even if a significant expansion into tourism occurs. It also follows that the strongly held opinions about the Aboriginal community in Jabiru will remain in the foreseeable future.

Administration and governance

A primary purpose of our research in Jabiru has been to examine the development authority model as a means of coordinating mining town development and administration. In this section we review several measures of resident involvement in community affairs and resident assessments of the Development Authority and the Advisory Council. Next, we turn to the possible agenda for the new Town Council and resident expectations of its likely effectiveness.

Although governance and administration had captured the interest of those interviewed in Darwin and Jabiru on early visits, we also formed the impression that decisions about who developed and managed the community were not riveting topics for most Jabiru residents. This indifference was the result, first, of the fact that residents had been given no say in the original decision to create a development authority and saw pressure to move towards an elected Town Council as based on the concerns of the Territory government and only a minority in the community. Second, mining towns tend to attract temporary residents who, in the words of one respondent, 'treat Jabiru as a short-term stopover [so] there's a lack of community feeling and involvement'. A third possibility was that regardless of the Development Authority, Jabiru was basically a company town and, according to another respondent, residents 'seem to believe they should have everything given to them without making an effort'.

The issue of 'apathy' and lack of 'community spirit', mentioned in chapter 5, was one which we wished to explore in the community survey, primarily to see if non-involvement was typical in Jabiru and if some parts of the community were less involved than others. Table 6.4 presents results from a wide range of questionnaire items. The first panel in the table shows that 14 per cent of respondents never 'discuss community problems or town administration with other people ', and approximately 30 per cent of each of the four analysis groups discuss community issues several times a week. Differences across the groups emerge when the average number of activities in which respondents take part is considered. The data show people in married households were more involved than singles in community activities and that Ranger households were more active than non-mining families. (Length of residence in town was, predictably, very strongly related to participation. Those who arrived in 1984 had done an average of 2.8 of the 12 activities compared, at the other extreme, to pre-1982 arrivals who averaged 4.9 activities.)

Equally interesting is the finding that in spite of the *average* number of activities, some residents were involved in none or only 1 of the 12 things we asked about. Thirty per cent of the non-Ranger married respondents fitted into that category compared to only seven per cent of Ranger marrieds. The average level of participation by non-Ranger marrieds was raised to almost four (3.9 activities) by the 13 per cent of their number who said they had done eight or more of the items on the list.

Most frequent types of involvement in the 12 possibilities nominated were: member of a social club (78 per cent), member of a sports team or sports club (60 per cent) and member of a union (49 per cent). It should be pointed out that although membership in these organisations would lead to a major commitment in time and energy for some, union membership for many residents was simply a means to employment, and belonging to a social club was the only way to buy alcohol legally in Jabiru.

For the other more 'discretionary' activities on the list, 44 per cent said that they had talked to a 'Development Authority, Advisory Council, Ranger or other official about a community issue', and 38 per cent had 'helped in the running of a school fête, community art show or other community activity'. Least frequent of the activities were being a candidate for the Advisory Council or the Town Council (only married people and only two per cent of all respondents) or an office holder in a union (seven per cent) and writing to a local official about a community issue (eight per cent).

The most notable contrasts across the columns in this part of the table were that married respondents helped with school fêtes and other community activities more than singles; that married people were

Table 6.4 Community involvement
(percentages of respondents)

	Ranger married	Non-Ranger married	Ranger single	Non-Ranger single	Total
Discussing community issues					
How often discuss community problems with others?					
Never	9	16	20	15	14
Once or twice a month or less	39	27	36	40	36
Once a week	21	23	12	16	19
Several times a week	31	34	32	29	31
Types of community involvement					
Member sports team/club	65	54	65	53	60
Member social club	89	71	70	76	78
Member church	15	17	13	15	15
Office holder in sports club, social club or church	34	28	22	22	27
Member union	45	36	66	48	49
Office holder in union	12	4	9	—	7
Helped run school fete, art show or other community activity	47	47	30	24	38
Organised school fete, art show or other community activity	25	27	16	26	24
Attended meeting focused on local government or other community matter	39	32	28	15	29
Been candidate for office in JTAC[a] or JTC[b]	3	4	—	—	2
Talked to local official about community issue	51	55	30	34	44
Written to local official about community issue	8	14	8	5	8
Number of above twelve activities done					
None	2	10	9	4	6
One	6	20	12	21	14
Eight or more	8	13	8	—	7
Average number of above twelve activities done	4.3	3.9	3.6	3.2	3.8
Town Council election					
Voted in JTC[b] election	32	28	12	16	23
Number of elected candidates identified by respondent					
None	22	23	50	54	36
One or two	24	27	29	23	26
Three or four	35	32	18	22	27
All five	19	18	3	1	11
Average number of candidates identified	2.6	2.4	1.1	1.2	1.9

Notes: a Jabiru Town Advisory Council
b Jabiru Town Council

more likely than singles to have attended a meeting where community matters were discussed and were even more likely to have talked to one or more local officials about a community issue; that married respondents were more likely than singles to hold office; and that Ranger respondents were somewhat more likely than others to report being a member of a sports team or sports club.

The final section of the table reinforces the impression that single members of the community were less involved than marrieds in community affairs. Only 23 per cent of all the respondents reported voting in the May 1984 Town Council election, but married residents were roughly twice as likely as the singles to have voted. Our survey was in the field during the three weeks immediately after the new Council took office and it is significant that over half the single population (and 36 per cent of the total sample) were unable to name even one of the newly elected town councillors. On the other hand, close to 20 per cent of the married respondents (both Ranger and non-Ranger) were able to name the five new councillors, a task beyond all but a few of the singles. (The elected candidates were established residents of Jabiru and three of those elected to the Town Council in May 1984 had been members of the Jabiru Town Advisory Council since 1982. Only one member of the new council was an employee of Ranger; he had also been a member of the Advisory Council. The non-Ranger councillors included a minister and the deputy head-master of the primary school (who had both been Advisory Council members) and two successful local businessmen.)

We expected that the longer people had lived in Jabiru the more likely that they would have been to vote and to identify the councillors elected, and that proved to be the case. Thirty-seven per cent of those who came to Jabiru before 1982 voted in the election compared to only four per cent of those who arrived in 1984. Seventy per cent of the recent arrivals knew none of the elected councillors by name, but it is interesting that that was the case for 20 per cent of the long-term residents as well.

We consider the respondents' expectations about the new Town Council later in this section, but at this point we turn to assessments of town administration and governance. The first two entries in the table refer to items asking residents to rate the openness and responsiveness of town administration by indicating whether they agreed or disagreed (and how strongly) with two statements in the questionnaire (see Table 6.5).

For both statements agree and disagree responses were almost equal. Half felt that town administration in Jabiru was 'pretty much above board' and half did not. Half said that there was little connection between their interests 'and what happens in this

Table 6.5 Assessments of the Jabiru Town Development Authority, the Jabiru Town Advisory Council and the Jabiru Town Council
(percentages of respondents)

	Ranger married	Non-Ranger married	Ranger single	Non-Ranger single	Total
Hardly anything to do with town administration in Jabiru takes place behind closed doors; everything is pretty much above board					
Agree strongly	14	9	4	1	8
Agree somewhat	42	43	44	45	43
Disagree somewhat	28	33	37	41	34
Disagree strongly	16	15	15	13	15
There doesn't seem to be much connection between what people like me want and what happens in this community					
Agree strongly	10	8	12	18	12
Agree somewhat	32	42	48	30	37
Disagree somewhat	43	40	30	45	40
Disagree strongly	15	10	10	7	11
Effectiveness of Jabiru Town Development Authority (JTDA)					
Very effective	21	30	18	27	24
Somewhat effective	50	40	26	25	36
Somewhat ineffective	10	6	9	9	9
Very ineffective	4	2	2	4	3
Don't know	15	22	45	35	28
Effectiveness of Jabiru Town Advisory Council (JTAC)					
Very effective	4	—	2	1	2
Somewhat effective	16	20	12	—	12
Somewhat ineffective	13	11	11	9	11
Very ineffective	4	5	—	1	3
Don't know	64	65	76	89	73
Expected effectiveness of the Jabiru Town Council (JTC)					
Very effective	28	8	8	2	14
Somewhat effective	39	38	25	27	33
Somewhat ineffective	4	7	5	8	5
Very ineffective	1	3	—	—	1
Don't know	28	44	62	63	47
Functions of the new Jabiru Town Council should be					
Improve town facilities	72	68	66	52	65
Run the town	20	20	19	13	18
Encourage community cooperation	10	16	12	13	12

community' and half disagreed. Differences across the columns are modest although the Ranger married residents were slightly more positive than the other groups about the openness of Jabiru's town administration (56 per cent agreed that 'everything is pretty much above board') and were more likely to disagree with the statement that there was not much connection between their wants 'and what happens in this community' (58 per cent disagreed). Single respondents and non-Ranger respondents were marginally more likely to think that Jabiru's administration took place 'behind closed doors', but those patterns did not recur with the second agree/disagree item. In that case although Ranger residents were more satisfied than non-Ranger among the married households, the reverse occurred among the single households.

The next panels of Table 6.5 focus on ratings of the effectiveness of the Development Authority and the Advisory Council. Two aspects of the results stand out. First, 28 per cent were unable to rate the Development Authority and 73 per cent were unable to rate the Advisory Council. In both cases singles were less likely than marrieds to provide a rating—up to 89 per cent of single non-Ranger respondents who were unable to assess the Advisory Council. The lack of recognition of the Advisory Council compared to the Development Authority was understandable given the former's more recent (only since 1982) and limited involvement in the community, but with regular attention given to both organisations in the *Jabiru Rag*, together with publicity given to the council elections only a few weeks before the survey, such high percentages of 'don't know's' were not expected. The second aspect of the table that stands out is the very positive assessment received by the Development Authority. Five out of six of those rating the Authority said that it had been effective (including one-third who said that it had been 'very effective'). In contrast, only half of the few who rated JTAC saw it as being effective and this included less than ten per cent who felt that it had been 'very effective'.

We found that many respondents confused the Advisory Council with the Development Authority when answering these questions. All of these ratings were probed, however, and from the probes it was possible to identify respondents who were apparently rating the wrong organisation. To remove as much of this source of error from the ratings as possible, this entire section of the interview was recoded by the authors. As a result, when ratings of the effectiveness of the Advisory Council appear to have been based on Development Authority activities (or vice versa), those ratings have been recoded as 'don't know'. Without this recoding the percentage coded as 'don't know' for the Advisory Council rating would have been 52 per cent rather than 73 per cent.

Residents were also asked how effective they expected the new Town Council to be. Almost half said that they did not know ('let's wait and see'; 'it's too early to tell'), a proportion which is not surprising given that the survey was carried out well before the new Council had completed its first month in office. Most of those who proffered a rating indicated that they did expect the Council to be effective (including close to one-third who expected it to be 'very effective'), which is a judgement close to that received by the Development Authority.

We received a wide range of suggestions when we asked residents what they would like the new Council to do, in part, we expect, because the Council had not yet established a record of what they could (or could not) accomplish. Up to three responses were coded per interview with about one in eight residents feeling that the Council should sponsor events to encourage people to become more actively involved in the community. Eighteen per cent made the obvious response, 'running the town', as the main task of the Council. Most replies, however, focused on ideas for new or improved facilities and services. In all, 65 per cent had at least one suggestion, with the main proposals being for improvements in the sporting area (from building squash courts to top dressing the ovals), playgrounds for young children and activities for teenagers, and facilities for tourists. A dozen or more residents also included the need to open a cinema, upgrade health and medical services, control dogs, improve shopping and increase commercial competition in Jabiru. Other specific ideas ranged from improving radio reception and upgrading the airport to building a high school and expanding the town library.

This study began by focusing on the role of the Development Authority, but we soon realised that town administration was inextricably connected to several associated issues including the National Park and Aboriginal ownership of land. It was interesting, therefore, to find that neither Kakadu, Aborigines nor town administration were mentioned very frequently among the important 'problems and issues' in the July 1984 survey even though subsequent parts of the questionnaire showed that some respondents held strong views on several of these topics. Instead, residents identified key issues in town as the local shopping facilities, social divisions between mine and other groups and between Ranger salaried staff and wage employees, lack of activities for teenagers, and concern about the Labor government's uranium policy. With the exception of the uranium issue, these concerns would be familiar in many of Australia's mining communities.

The final section in this chapter summarised a wealth of

information on resident participation in the community and on evaluations of town administration. The most striking aspect was the low involvement of single residents in town affairs. Predictably, length of residence was also an important indicator of participation, with 1984 arrivals being much less involved in community activities than 'oldtimers' who had been in Jabiru since 1981.

Obtaining an assessment of the performance of the Development Authority was one of the main purposes of the community survey. Despite some reservations and a sizeable 'don't know' response, overall resident evaluation has clearly been positive and the Authority was rated effective by 60 per cent of all respondents (and by 80 per cent of those who were willing to rate it). Many respondents remarked on the neat and tidy appearance of the town centre and public areas. From their viewpoint an attractive community was a good indication of effective town administration. Other residents commented on the range of facilities and services provided in Jabiru (even though most could usually suggest improvements) as evidence that the Development Authority had done its job well.

Discussion of Jabiru's future is continued in the final chapter, but it is evident that there is no dearth of ideas about what the new Town Council might do to improve the community. Although almost half of the respondents were unwilling to guess the likely effectiveness of the new Town Council, most of those who did thought that it *would* be effective. The main reasons given for this were that the new councillors were competent people able to cope with administration, that the new Council would have more power to get things done, and that the town already ran well and would probably continue to do so. Among those who were less sanguine about the effectiveness of the Council, a few were worried about a lack of local government experience among the new councillors, but the main reason given was that it would still not have the power needed to be effective. Respondents ascribed this lack of power to the influence on the town of Ranger, National Parks and Wildlife, the Aboriginal population, and the Northern Territory government, or some combination of these. These assessments from our second layer of analysis complement the policy-oriented information in earlier chapters and contribute directly to the concluding discussion.

7 Conclusions

...it is clear that the town and its people have over the years subtly extended their range of influence and have broadened the notion of their perceived rights (Palmer, 1985: 51).

It was suggested in the Introduction that the human and institutional response to development tensions and opportunities in the Uranium Province provides the context for our study of Jabiru. The rapid pace of change, much of it externally induced, and endemic uncertainties in the uranium industry are far from optimal conditions for the establishment of a new community. Aborigines and white settlers alike have been subjected to a range of pressures on their lifestyles of a kind which most metropolitan Australians would find intolerable. This volatile situation shows no signs of abating in the foreseeable future and our conclusions can only be of an interim nature.

We have deliberately avoided making moral judgements here on the rights or wrongs of uranium mining in the Alligator Rivers Region, or anywhere else, and have accepted the reality of its existence in our investigations. The Northern Territory debate in this respect has more to do with *who benefits* from uranium as a local source of wealth than questions whether it should be mined at all. Rural Australians have few opportunities to determine if and where major resource and infrastructure investment occurs but are in a position to influence some of the distributional consequences. It is not really surprising, for example, that Aboriginal groups in the Territory should use these limited opportunities to call for improved royalties from the uranium mining industry and use part of what they receive, as the Gagudju Association is doing, to invest in new developments in the Uranium Province.

Most of the book, however, has dealt with planning and governance issues in Australian mining towns and the particular experience of the new residents of Jabiru. A bi-level analysis was chosen for the case study which distinguished certain development choices which were external to events in the Uranium Province from internally generated

158

responses to occurrences in the early years after the town was completed. The latter exercise also enabled us to describe the population of Jabiru and many aspects of town life and provided a basis for some direct comparisons with research findings in other mining communities. We began the study by asking why the novel idea of a development authority was chosen to build Jabiru rather than the more familiar models of mining town establishment, and why it was replaced so soon by a form of elected local government. Answers might have been forthcoming through the exclusive use of political economy notions of ruling class co-optation of the working class in the era of multinational mining groups and such approaches have been followed with some success elsewhere (Bradbury, 1977; Thompson, 1981). We chose instead to look closely at the pattern of inter and intragovernmental relations which characterise the Australian federal system and, in particular, the aspirations of the new Territory administration which took over from the common- wealth in 1978. In addition, we paid more attention to a description of the community and responses to events in Jabiru than is usual in political economy studies of urban development. New Australian mining towns exhibit considerable differences in their social profiles from stereotypes derived from Europe.

Our knowledge of mining towns such as Nhulunbuy and Jabiru led us to hypothesise that the push for the introduction of local government in these communities could not be satisfactorily explained on the basis of their size or rate of growth alone. It seemed to be connected with the desire of the Territory government to be more closely involved in the future control of these rich resource regions than it had with the wishes of the majority of residents. This supposition suggests that there is rather more to the introduction of local democracy Territory-style than the creation of some autonomy in small centres outside Darwin. Achieving local government under such circumstances is to acquire a platform for popular expression of a kind which can be directly influenced by Territory-wide politics.

Governance choices

The initial question about a choice of a statutory development authority to establish and manage Jabiru was significant because it was the first time such an institutional model had been used for the purpose in an Australian resources project. There seem to be obvious advantages for Territory government and the mining companies in shared control of such an authority. The former could retain overall control of physical development (subject to the Parks Service town plan) while paying for only a minor share of the new town. The mining groups would save little financially from the arrangement but were shielded from open association with non-production and social overhead aspects of town development, so often the cause of industrial disputes in other resource projects. Yet, behind the scenes, they were still able to exercise enormous influence through control over most of the construction budget.

There were also important consequences for other interest groups in the Uranium Province. In the Fox Inquiry the commonwealth identified a clear responsibility towards the welfare of local Aboriginal peoples, conservation of the environment and control over the mining and export of uranium. It would not, therefore, acquiesce in any arrangement which did not provide adequate safeguards for these interests, none of which were suitably covered in other models of town establishment. A statutory authority answerable to conditions laid down in a head lease granted by the National Parks Service, and run by Northern Territory government service departments and the mining companies seemed an excellent compromise.

A further requirement which pointed to creating a novel institutional arrangement was the likelihood that Jabiru would be required to service the sequential needs of three mining companies. The possibility that more than one town would be built was never seriously pursued and meant that the awkward problem of retrospective cost sharing among the various participants would be necessary. We have shown that the inability of two companies to gain commonwealth approval to commence operations led to costs being initially covered by Ranger and government on the basis that headworks and certain communal facility costs would eventually be shared. This would be by means of reimbursement *pro rata* according to a formula related to the share of accommodation provided in the town.

Although the final cost sharing between ERA and the two governments had not been finalised when this book was written in early 1985, the agreement was close to approval. After the repayment of certain 'overpayments' in their shares have been made to ERA and the Development Authority, it seems that the Authority will be left with a debt of some $6.9 m comprising borrowings of $5.8 m and the accrued interest (personal communication: Geoff Stolz, Chairman, Jabiru Town Development Authority, 1985). These funds largely relate to the 144 serviced and vacant residential allotments which were not taken up by Pancontinental Mining and Denison Australia and are a clear indication of the financial risks involved in planning for multi-company occupancy in a resources town. The debt and the rapid accumulation of interest charges presents a major burden for the Development Authority as its only income (other than from the small number of ratepayers shown in Table 4.2) must come from charges levied on new businesses which might wish to establish in Jabiru and share in its public facilities. ERA has made a concession to the effect that all such monies should go to the Authority alone rather than being shared between them.

There is considerable uncertainty whether the new Plan of Management for the Kakadu National Park under preparation by the Parks Service will relax some development controls over Jabiru to allow diversification to occur. It is hard to visualise how small single industry towns like Nhulunbuy and Jabiru can achieve a secure economic future without a change of this kind and the necessary involvement of the private sector. The experience of mining town development in Australia suggests there will be little private sector investment in tourism or anything else in Jabiru under the leasehold tenure arrangements presently available. At the same time it can be argued that these circumstances provide a unique opportunity for Aboriginal involvement of the kind which has begun at the Cooinda complex in the Park.

The second of our questions queried the speed with which the Territory government sought to introduce local government to Jabiru and requires an assessment of several motives. We indicated there were clear reasons for the promotion of local government in the Territory based on factors such as improving administration, building up local responsibility, and devolving power as an aid to economic decentralisation. Heatley (1978) suggested that the rapid growth of small centres outside Darwin since the mid-1960s led to dissatisfaction with existing town management boards and to demands for local government in places like Katherine and Tennant Creek. This trend began before self-government and was extended by successive Everingham governments to the introduction of community councils in Aboriginal Communities (*Local Government Amendment Bill*, 1981), new towns (successfully in Jabiru and Palmerston but not in Nhulunbuy) and other small communities. There was a determined effort to devolve some power from Darwin for a variety of reasons (*NT Parliamentary Record*, Part 1, 16 March 1982: 2127). It was central to the Everingham political philosophy to draw rural communities into the decision making process. 'We are trying to build up responsibility in these people; we are trying to build up responsibility in municipal councils generally...' (*NT Parliamentary Record*, Part 1, 2 June 1982: 2439-40). In addition there was the conviction that the Territory had to decentralise economic opportunities with the key being people-led initiatives at the '...third tier of government—city and shire councils' (Paul Everingham in *The Australian*, 17 January 1985).

An interesting facet of this strategy, clearly demonstrated in Jabiru, was its Darwin-inspired character and its marginal relevance to the feelings of a majority of residents. This situation is not unusual in company towns and it is evident that the motivating force for change in Jabiru came largely from people outside the mining industry. In the case of Nhulunbuy the strategy proved unsuccessful: '...some years ago we offered local government to the people of Nhulunbuy. Unfortunately the people of Nhulunbuy seem to have spurned that offer. There has been no visible movement or agitation towards local government there that I have been able to discern in 3 years' (Paul Everingham, *NT Parliamentary Record*, Part 2, 19 October 1983: 366). The reasons for this are hard to establish without separate investigation and persist in spite of considerable dissatisfaction with certain aspects of company control in the town (Lea, 1984).

It has been our contention throughout this study that there was a hidden political agenda which was not expressed publicly and is likely to have fuelled the Territory government's desire to have local government in place in Jabiru as soon as possible. We suggested there was a firm conviction in CLP circles that a likely avenue for increased Territory involvement in developments in the Uranium Province was to provide a platform for the expression of local opinion through elected local government. This was a bi-partisan attitude in Darwin with Government and Opposition in favour of it but for somewhat different reasons. The Everingham government believed in promoting more uranium mines and the creation of an open and diversified regional centre to serve them. The Opposition adhered to the Fox Inquiry controls but welcomed the principle of locally elected representation within these overall constraints (*NT Parliamentary Record*, Part 1, 7 June 1984: 512-13).

In political terms the CLP probably stood to gain considerably from the early introduction of local government in Jabiru if this was to assist the process of diversification and its transformation into something more versatile than a 'closed' mining town. The development was feared by Aboriginal organisations and one reaction is illustrated by the approach of the Northern Land Council to the pressures for expansion of the tourist industry in Kakadu (Palmer, 1985). Local government was the thin end of a very broad wedge if its presence was to herald substantial political and economic changes in the Uranium Province and set a precedent for a similar process to occur in the Gove Peninsula. The Gagudju Association seemed less apprehensive of the changes so long as it was able to secure its own economic objectives in the town. The likelihood that the Association might secure rights to build and manage a motel in Jabiru arose in early 1985 and is a possibility which removes much of the political objection in commonwealth circles to changes in the economic base of the town.

It is interesting to observe that Aboriginal sources have been apt to blame the town and its people for 'subtly' extending their range of influence and range of perceived rights (Palmer, 1985), when the origins of this process were in Darwin rather than the Uranium Province. It is extremely unlikely that local government in Jabiru would be in existence today if its presence had been reliant on citizens' petitions of the kind pioneered in Castlemaine and Ballarat a hundred years ago.

Planning models for new resource-based communities

There are few, if any, examples in the Australian urban studies literature of attempts to categorise or identify the planning models employed in the various phases of resource-based community development. Different stages of project development demand appropriate social planning frameworks tailored to the successive priorities of the emerging community. This idea of deliberately *changing* the emphasis of planning activity as the early feasibility phase of the project moves to implementation procedures and, finally, to development itself and control by local democratic institutions is well established in the Canadian literature. The notion advanced by Paget and Rabnett (1983) that planning must be involved with the design and encouragement of governance institutions to the same degree as the familiar physical and economic considerations is a major departure from Australian experience in the past.

Based on their experience of developing the new coal town of Tumbler Ridge in northeast British Columbia, Paget and Rabnett (1983) illustrate three broad changes in the planning sequence in a large resources project. There is an initial phase of *socially sensitive planning* which is an amalgam of social impact assessment 'with its emphasis on social objectives and its concern with making socially sensitive decisions in which potential repercussions are considered', and planning activity which reflects 'a concern with designing creative solutions to the social problems of communities...' (Paget and Rabnett, 1983: 8). As they point out, however, a major difficulty in new projects like Tumbler Ridge is the absence of any existing community and local institutions, which requires the developers to hold the interests of the future community 'in trust'. Thus the first phase of project development is inevitably technocratic or 'top-down' in approach. Surrogates for the future residents are found in the study of older communities in relatively similar circumstances.

In the case of Jabiru and the Uranium Province this initial planning phase markedly failed to recognise the presence let alone the needs of existing Aboriginal groups except in a cursory manner. It was an omission redressed to a considerable extent by the Fox Inquiry recommendations arising out of the first use of the commonwealth's environmental impact legislation. Planning activity subsequently paralleled the impact analysis in a very similar fashion to the Canadian example.

The second phase or implementation stage of the resources project requires a change in planning approach to *planning by invitation*. Here the emphasis shifts to 'building of bridges between organizations, involving those who will ultimately own the community's problems.... It emphasizes governance/stewardship' (Paget and Rabnett, 1983: 14). In British Columbia a local government approach is seen as superior to other implementation possibilities because it is thought to be more sensitive to people and better enables political roles (governance) to be balanced against technical roles (project management). The local government District of Tumbler Ridge was incorporated at the outset and a Commissioner appointed who would act as Mayor and Council to manage the transition to fully elected local government.

The experience of building Jabiru followed a remarkably similar path under the guidance of the Chairman and General Manager of the Jabiru Town Development Authority though the question of governance transfer was a good deal less certain than in the Canadian example. Jabiru was not incorporated as a local government area from the beginning but the creation of the interim Advisory Council had a similar effect. There was a major difference, however, in the control mechanisms over land in the two towns. In the Canadian case the Province was to provide the new municipality with a five-year supply of Crown land at 'wilderness prices' to facilitate affordable housing, diversified industrial employment opportunities and other important communal facilities (Paget and Rabnett, 1983: 17). In Jabiru, of course, title to surrounding Crown land was granted to local Aborigines and the town site itself vested in the National Parks Service. Although this arrangement was to prove extremely effective in conserving the natural environment of the region, it presented major obstacles in achieving 'normal' local government and possible diversification in the future. The latter effect is hardly surprising as it was the declared intention of the Fox commissioners to do just this. The fact remains, however, that the 'solution' devised by the Fox Inquiry incorporated some mutually exclusive objectives for the town when the recommendations are reconsidered. It seems hardly feasible in hindsight to allow the development of a $100 m town in the middle of a major tourist attraction like the Kakadu National Park and expect the residents to remain as 'visitors' given the sort of pre-conditions which are necessary before local government can take root under Australian conditions. As we saw in chapter 2 one of these is the private ownership of land. No mention of governance arrangements in the town was included in the Fox Inquiry Report (Fox et al., 1977). The notion of a town of temporary residents was at variance with the strongly held belief in the Territory that every effort should be made to establish permanent settlement on the north Australian frontier. Under these conditions it is not surprising that initial constraints on the development of Jabiru were seen in some quarters as little more than a temporary obstacle.

The final development phase of the new town begins with the arrival of the first residents and overlaps the preceding implementation period for several years. The emphasis is now on 'community development' and the planning process changes from design and establishment activities to the *building of a new community* (Paget and Rabnett, 1983: 21). Central to this process is the appearance of local self-government and the 'client' for ongoing planning activity becomes the community itself. This phase in the early life of a resource-based community is not susceptible to top-down forms of social engineering and attempts to introduce popular participation may easily meet with the sort of opposition found in the initial meetings at Jabiru. This experience points to a disproportionate involvement of those employed outside the mining industry for a variety of work-related reasons. Even though the latter do not figure prominently in election manifestos published in the *Jabiru Rag*, there are numerous reasons why the non-mining minority in Jabiru should view the Town Council as their only means of expressing their opinions on community issues to government and the mining company. Mine employees can achieve the same objective within their unions and the company.

Planners involved with remote resource towns have a clear obligation to identify and understand the special social characteristics of these places if they are to have any relevance to the community establishment process during the early years. Some previous contributions to furthering such understanding were summarised in chapter 1 and covered a range of descriptive and explanatory attempts to model social change and development in mining society. Among them was Bulmer's (1975) list of eight ideal characteristics of the European (mainly English) mining community which we will now contrast here with some of the results from the Jabiru study.

Some social characteristics of the new Australian mining town

One by-product of the community survey was the large amount of descriptive social data enabling us to draw some preliminary comparisons between Bulmer's ideal characteristics and the circumstances of a new Australian mining town. We acknowledge, however, that to do this job properly demands a study directed to answering questions raised in Bulmer's generalisations and this would have meant a focus on Jabiru society rather than governance. Nevertheless, the generalisations draw attention to areas of community life in Jabiru which differ from the European norm even if they do not set out to explain them.

Physical isolation

This well known characteristic of Australian rural settlement shows up in Jabiru in the concern of residents over requirements for better basic services in areas such as health care, retailing and entertainment. Interestingly, Jabiru does not appear to be isolated enough. Three hour access to Darwin on a good sealed highway places the town within the metropolitan catchment for major services and, paradoxically, reduces the chances of improving the situation in the foreseeable future. For Jabiru the crucial issue is size rather than isolation.

Economic predominance of mining

Subservience of settlement or village to the mine is held to be a standard characteristic and Jabiru is no exception in this respect. Some of the effects are less extreme than in other isolated mining towns in the Territory because of the presence of two major commonwealth instrumentalities, the Office of the Supervising Scientist and the National Parks Service. They provide limited diversity in the employment market and respondents from these groups gave articulate responses to the survey. One of them when asked about Jabiru as a place to live said, 'It's a concentration camp for the mining population'.

Nature of work

We collected little information which pointed to the effects of mining employment itself on community life. In the European stereotype the dangers, skills and long apprenticeship associated with underground mining extends into the social fabric of the village. New Australian mines in the north are mainly open pit and characterised by very different working conditions. Few wages staff require a lengthy period of training and there is little immediate danger in the workplace. That is not to say, however, that working in a uranium mine is free of long-term dangers of radiation or whether it might generate other effects connected with the controversial nature of the industry.

Social consequences of occupational homogeneity

The notion of a society stratified according to the position of the male household head in the mine workforce with few opportunities for upward social mobility persists in the responses of some residents. A typical one being, there is 'a line down the middle of the town between staff and wages—shows in the JSSC (Social Club) and the Golf Club'; or, a 'division between award and staff...mainly the wives and not the men'. It is likely that the effects of this division on upward social mobility are strongly conditioned in north Australia by high levels of voluntary or company-induced turnover in the workforce. This is a subject which requires separate investigation and longitudinal analysis over several years.

Leisure activities

In the European stereotype is the suggestion that leisure activity in the mining community is designed to contrast with the constraints of underground work and that occupational ties are actually reinforced in outside activities. Although our data were not collected in a way which can confirm or reject these observations, there is a considerable amount of information about behavioural activity out of working hours. Jabiru's unique location in a national park presents opportunities and constraints over the use of leisure time which are not shared to the same degree elsewhere. It is not unusual, however, for mining town residents in north Australia to suffer considerable restrictions over outdoor activities, most of which stem from the presence of surrounding Aboriginal lands. Nhulunbuy and Nabarlek in Arnhem Land and Alyangula on Groote Eylandt share this characteristic and their occupants are required to observe strict controls over leisure activity beyond the boundary of the town or mining lease.

The family

The notions of strict segregation between male and female roles in the mining family with the former concentrating on work outside the home and the latter on things within it, contained in the Bulmer (1975) summary, demands more detailed consideration than can be achieved here. The Jabiru employment data, however, reveal that only one-third of female household heads were occupied by home duties alone. The majority of women who had jobs were employed in similar proportions by the mine, private sector, Northern Territory government and a miscellaneous group of other employers. Only one-third of women had fulltime jobs, however, and there were suggestions among the survey responses that some of those on part-time would like something better. In overall terms the pattern of females in the workforce is not dissimilar to the national norm. Marital breakdown and other aspects of family life in Jabiru were suggested to us as being exacerbated by living in a community of this kind but the causal relationships are not easy to pin down and many may not be related to mining at all.

Economic and political conflict

There is no doubt that Jabiru has suffered from its fair share of industrial conflict in its short history and has a comparable record in this respect to other mining centres in the north. An important point made by Bulmer (1975) is that opposition to the company in the workplace in mining towns can be met by the denial (or conditional granting) of various basic domestic and social needs. Thus company controls over subsidised housing are a powerful means of responding to industrial demands at the mine. There is direct evidence that this weapon was used to some effect during the lengthy 1983/84 strike in Jabiru and is a disincentive to the more extreme forms of union action so long as no alternative tenure forms exist. It could thus be argued, counter to the views expressed by some political economists, that the introduction of employee home ownership schemes and the building of public housing in mining towns like Newman is actually in the long-term interests of the labour unions by lessening dependence on company housing as long as alternative forms of employment are available.

The whole

There is a suggestion that something special about community life in mining towns stems from the interrelationship of all the characteristics noted here. This may still be the case in European mining villages but if the same phenomenon exists in the modern Australian examples we have discussed in this book, it is certainly very different. The single constant feature underlying so much of community and family life in Jabiru has been the presence of uncertainty and it is this, more than anything else, which seems to determine patterns of behaviour. For most, going to north Australia is a temporary move and working in Jabiru is probably one of several options open to those attracted to the special economic advantages of mining life. Even so, the survey showed that the family incomes of approximately half the households were under $27 500, hardly a passport to quick riches. Responses to questioning about whether they should move or stay showed that more than half planned to leave Jabiru in the next two to three years. Almost half the respondents owned a house and/or land elsewhere with Queensland proving to be the most popular location. Such facts are a strong indication that most people in Jabiru tend to see themselves as visitors in the Uranium Province even though they are anxious to acquire the sort of rights which accompany permanent residence in a community. To this extent the Aboriginal organisations are correct, Jabiru is largely a visitors' town (Palmer, 1985).

These considerations lead naturally to a discussion about the future of the town and forms the final section of our conclusions. Any attempts to forecast future events in the Uranium Province are fraught with difficulty but we have nonetheless considered it important to seek the views of those living in the town and who have formed the closest links with it.

Jabiru and the future

Jabiru's overwhelming dependence on uranium mining is a reality little affected by talk of new tourism developments in the town and Park. Community perceptions of the town and its future are thus inextricably linked to attitudes towards the uranium industry itself. When residents were asked how they felt about uranium mining in the survey they were given a fourfold choice between: leave it in the ground; use it for medicine, science and electric power only; use it for the foregoing plus military uses; and do not know. Most chose the second option accounting for almost 80 per cent of the total, only seven per cent suggesting it should be left in the ground and 10 per cent including the military use as well. There was little variation across the household categories though marginally fewer Ranger marrieds were against mining than the others.

When it came to the question about Jabiru's future as a mining town the most popular guess was it would last between 10 and 25 years (39 per cent). closely followed by 31 per cent who thought 25 to 50 years more likely. Only four per cent seemed to believe the widely reported threats about mine closure were accurate and opted for less than 10 years. A closely related question inquired if Jabiru would survive after mining ceased and here half the respondents thought it would (51 per cent) and 31 per cent that it would not. Although these responses have no bearing upon what may actually happen in the future they are nevertheless an interesting measure of community uncertainty. Jabiru and uranium are indivisible for the present population of mining and civil service households but an alternative tourism-based economy would probably take over were the mine to close. However unpalatable this eventuality might be for a development oriented Northern Territory goverment or for local Aboriginal organisations reliant on royalties, it does at least offer a future diversity absent in most other resource-based communities in the north.

The antipathy between environmental conservation and mining, so vigorously expressed in the Fox Inquiry hearings, has left a choice of economic options which will probably ensure the town's survival. Less easy to predict is the effect of Aboriginal involvement on town and region. Once again it is a rich paradox that the most politically significant reason which will be found in Canberra for continuing uranium mining in the north is the importance of Aboriginal royalty payments. Moves are already underway from the Aboriginal side to renegotiate the Ranger Agreement (*The Australian* 27 July 1985) and few would deny that this is a much easier option for all concerned than closure of the industry.

The appearance of local government in Jabiru has been portrayed as externally induced and accepted by a less than enthusiastic community, most of whom had little interest in it, an attitude reflecting the tightly controlled and generous conditions of modern mining life in the north. It is a situation re-enacted on many occasions in a century of Australian mineral exploitation but the contrast between Jabiru in 1984 and the scenes at the first municipal elections in Hill End, New South Wales, as described by David Kennedy Jnr in the 1870s, could scarcely be greater. It is a fitting note on which to end this book:

> The town is excited over its first municipal election...We find the chief street closely packed with human beings eagerly expecting the appearance of the candidate on the hustings...The first, crushing his felt hat under his arm with the earnestness of his speech, vows he will do his utmost to economise the public funds. 'Hurrah.'
>
> The second, almost twisting the buttons off his coat, declares he will go in for free education. 'Good for you!' The third, running both hands nervously through his hair, announces his fixed intention of devoting himself to the mining interest. 'Go to it old chap!'
>
> The fourth, with clenched uplifted fists, denounces vehemently the opposition he received from the despicable Teetotal, Masonic, Good Templars', and Odd Fellows Societies. At this there is a tremendous uproar, and those in front make loud slighting remarks as to the speaker's parentage and fitness for aldermanic honours (Reproduced in Keesing, 1971: 283-4).

Appendix A
Selected writings on the Alligator Rivers Region

1 Aborigines, pre-history, exploration and early settlement

Leichhardt (1847)
Spencer (1928)
Warburton (1934)
Berndt and Berndt (1954)
Bauer (1964)
White (1967)
Cole (1975)
Barker (1978)
Keen (1980)
Levitus (1982)

2 Uranium discoveries

The Australian Financial Review 27 May 1954
Northern Territory News 18 and 22 August 1970
Annabell (1971)
Southern Miner 2 November 1970
Ryan (1972)
The Australian Financial Review 16–20 July 1973

3 Ranger Uranium Environmental Inquiry

Fox et al. (1976; 1977)
Commonwealth of Australia (1977)
Woods et al. (1978)
Harris (1980)
Saddler (1978; 1980)
Formby (1981)
Saddler and Kelly (1983)
Roberts (1984)

4 Impact on the Aboriginal community

Parsons (1978)
Coombs (1980)
Roberts (1981)
Stanley (1982)
Tatz (1982)
Tatz et al. (1982)
Von Sturmer (1982)
Altman (1983)
Howitt and Douglas (1983)
AIAS (1984)

5 Kakadu National Park and tourism

Christian and Aldrick (1977)
Ovington and Boden (1979)
AIAS (1980)
ANPWS (1980)
Fox (1982; 1983)
Peat Marwick Mitchell Services (1982)
Forbes and Merrill (1983)
Gale and Jacobs (1984)
Sullivan (1984)
Weaver (1984)
Palmer (1985)

6 Town of Jabiru and the Ranger Project

Commonwealth of Australia (1973)
A. A. Heath and Partners (1973)
Cameron McNamara et al. (nd)
A. A. Heath and Partners (1978)
Cameron McNamara-Minenco Joint Venture (1979a; b; c; d; e; f)
Woods (1979)
Simpson (1980)
Bath et al. (1980)
Young (1983)
Stolz (1983)
Danielson (1984)
Lea (1984)
Bath (1984)
Zehner and Lea (1984)
Lea and Zehner (1985)

Appendix B Survey methodology

Survey sampling

Carrying out a home interview survey of residents in a community involves several stages. Initial decisions have to be made about the number of cases needed for analyses and, because it is generally too costly and unnecessary statistically to interview everyone in a community, about the most appropriate sampling design. For the Jabiru Community Survey a decision was made to aim for 100 Ranger married, 100 non-Ranger married, 50 Ranger single and 25 non-Ranger single interviews. These categories reflected the main factors which were thought to distinguish residents and were chosen to provide a structure for subsequent analyses. The emphasis on married households was a consequence of our particular interest in governance and administration and an understanding, later confirmed by the survey results, that singles in Jabiru were less involved in governance issues and therefore less likely to provide assessments of administrative arrangements in the town.

The decision to include single people in the sample was made despite a general reluctance of researchers to include them in studies of mining towns. Not including singles has been done primarily because they are harder to locate than married residents (something which we also found to be the case) and/or because their problems have in the past been seen by some observers as 'less central to the main theme' of living in remote communities (Brealey, 1972: 13). As a consequence, one of the most productive of the research organisations studying mining towns in Australia, CSIRO's Resource Communities Environment Unit, has focused almost exclusively on 'married women' or 'spouses' in some 13 surveys they have undertaken between 1971 and 1982 (Neil, 1982:33).

Excluding single residents was not seriously considered for the Jabiru survey for three reasons: first, we wished to determine the extent to which ideas and reactions of single people—often a very high proportion of the households in a mining community (over 40 per cent

173

in Jabiru)—differed from those of married residents; second, we wanted to gauge the success of the dispersion of single persons quarters (SPQs) in residential neighbourhoods as a technique for handling the housing of single people—and for that we needed to talk to singles as well as their married neighbours; and finally, we wanted to come to conclusions about how Jabiru worked as a community and how its residents perceived the transition toward local governance. There is no accurate way of doing those things without including the single population in the assessments.

For sampling purposes, it was obviously very helpful that in a large majority of cases the Ranger/non-Ranger and marital status of households could be determined on the basis of the type of housing in which people lived. Ranger SPQs housed single Ranger employees; government SPQs accommodated single non-Ranger residents and so forth. However, while most dwellings in the caravan park and the housing remaining in Jabiru East were occupied by non-Ranger households, the marital status of those residents could not be determined beforehand. As a result, five housing-related strata in the community were enumerated for sampling purposes (see Table B.1 below). It was clear that if we wanted to obtain the same number of interviews from each of the two groups of married households, a higher proportion of non-Ranger addresses had to be included because there were many fewer non-Ranger than Ranger married addresses in the community. Non-Ranger marrieds thus had a much better chance of being included in the sample than Ranger marrieds.

A stratified systematic random sample of addresses was selected from lists which were compiled from housing data made available by the Development Authority and Ranger. The sex of the respondent was specified randomly for each address as part of the sampling procedure to obtain roughly equal numbers of male and female respondents in married households.

As the fieldwork progressed several things became apparent. First, approximately 20 per cent (much higher than expected) of the non-Ranger married housing addresses were either vacant or in use only as motel-type accommodation for government employees who were working temporarily in Jabiru or the surrounding area. Such addresses, when encountered, were not included in the survey. Second, the lack of alternative accommodation led to single parents living in houses and to other living arrangements that had not been foreseen when the sampling was done. These anomalies forced some adjustments in the proportions of addresses sampled within the five sampling strata, the main change being that a larger number of married non-Ranger addresses were included. Careful records of the sample were kept, by strata, and as a result it was possible to weight and combine interviews to calculate percentages that represent the households in Jabiru as a whole.

A modified version of the survey questionnaire was also prepared for use in the Manaburduma Aboriginal Camp. The questionnaires were administered by representatives of National Parks and the Gagadju Association, but only two of the six or seven Aboriginal family groups in the camp were interviewed and it proved impossible to contact other families during the 'dry' to secure a more representative response. The questionnaire was difficult to administer even in modified form, however, and the two sets of responses obtained are not included in analyses reported here.

Fieldwork and response rates

People with whom we had contact within the Development Authority, the Advisory Council and Ranger were aware several months in advance that a survey was planned for July 1984. Just prior to the start of the interviewing an item in the local *Jabiru Rag* announced the survey and 'introduced' us to the community as researchers and interviewers. The article noted that the study had the support of the Development Authority and it was hoped that the new Town Council would benefit from having access to the results.

Some prospective respondents were suspicious of the study despite the item in the *Jabiru Rag*, and Ranger was seen as possibly behind it all. During the first week of interviewing we heard of mine workers asking union representatives if they should allow themselves (or their spouses) to be interviewed. As it turned out, two of the early respondents were key union officials who apparently felt, after answering the questionnaire items themselves, that the information could be of use to the community and that the study was not trying to push a pro-Ranger (or pro-union) barrow. Their advice was, reportedly, that union members could take part or not as they wished. Had the decision been to discourage participation, there is no doubt that the rate of response and the generalisability of the eventual survey results would have suffered markedly.

In the event, the three weeks of interviewing worked out exceptionally well due partly to the perseverance of interviewers, but primarily to the generosity and hospitality of residents. Each of the interviewing targets was met or significantly exceeded: Ranger married yielded 118 interviews; non-Ranger married, 100 interviews; Ranger single, 64 interviews; and non-Ranger single, 41 interviews. The interviews averaged 54 minutes; slightly longer for married respondents and slightly shorter for singles.

Perhaps the most rewarding aspect of the fieldwork was that the response rate (number of interviews divided by the number of occupied addresses) for the study as a whole was 89 per cent, a rate of response that would be among the highest obtained in a study of this size which included all sections of a community in its sample. Table B.1 summarises sampling and response information for the survey. It is worth remembering that living arrangements for residents did not always correspond to sampling categories. Several government houses, for example, were occupied by single parents and 'exceptions' occurred in each strata. Most interviews generated by non-Ranger addresses in Jabiru East and the caravan park turned out to be with married households.

Table B.1 Sample strata and response rates

Sample strata addresses	Occupied dwellings	Number in sample	Number of interviews	Response rate %
Ranger married housing	207	126	116	92%
Non-Ranger married housing	84	79	75	95%
Ranger SPQs & demountables	141	76	69	91%
Non-Ranger SPQs & demountables	123	34	22	65%
Jabiru East; caravan park	52	49	41	84%
	607	364	323	89%

Refusals were encountered at only three per cent of the addresses and some remaining cases of non-response occurred because residents were on holiday. The compact size of the Jabiru townsite was one reason for the high rate of response. Because over half of the addresses in the community were eventually included in the sample, interviewers were seldom far from the next sample address. This not only reduced interviewer travel time within the townsite to a minimum, but also made it easy to check back on 'not at home' respondents. The intent had been to limit these call backs to six per address. In practice it was not at all difficult to check on dwellings more often as we criss-crossed the community. Ranger SPQs were especially easy to check because they were on the pathway system and convenient to the mess hall where interviewers had their meals. Lower response rates for the non-Ranger SPQs are undoubtedly related to their somewhat greater dispersion around the townsite (see Map 14).

It seems likely that another reason for success among Ranger singles was that prospective respondents were able to look *us* over in the mess hall, and several people volunteered to be interviewed even though they had not been selected in the sample. Although over half of the interviewing was eventually done by the authors, the fact that two of the interviewers were young women did not hurt the response rate among single males.

Sampling error

Even in a properly conducted sample interview, survey results are subject to several types of error. These include non-response errors, errors in reporting and processing, and sampling errors. Because of the high rate of participation attained in this survey, we do not expect that the data set suffers from any serious biases due to non-response. Similarly, we trust that close supervision and/or direct involvement by the authors in the interviewing, coding, and data entry stages of the survey operation have reduced any errors that could occur in those activities to negligible levels.

Estimates of sampling error reflect the fact that responses from a sample are very unlikely to *exactly* mirror those from the entire population. For example, although 55 per cent of a population prefer Candidate A, a *sample* taken from that population might indicate that 60 per cent (or 50 per cent) prefer Candidate A. This variation or 'error' can be reduced by increasing sample size. The primary determinant of sampling error is the absolute number in the sample, but this type of error also decreases when the sample represents a high proportion of the total population.

The entries in Table B.2 are equal to two standard errors and allow one to infer (with 95 per cent confidence) that a percentage from the sample survey, plus or minus the sampling error, accurately represents the population from which the sample was taken. These figures take into account the numbers of interviews obtained and have been adjusted to reflect the particularly intensive sampling which was done for this study. It is interesting to note that the estimated sampling errors for the non-Ranger married group, which was very heavily sampled, are virtually the same as those for the study as a whole.

For those unfamiliar with sampling errors, an example might clarify how the percentages provided in Table B.2 can be used. In chapter 6 (Table 6.2) we found that 50 per cent of the residents of Jabiru reported that one disadvantage of Kakadu was the limits placed on what people could do and where they could go within the National Park. The entry in the 'Total' column and 50 per cent row of Table B.2 shows a sampling error of 3.8 per cent. This indicates that we can be 95 per cent confident that a survey of all the households in Jabiru (not only the ones in our sample) would find that 50 per cent of the community, *plus or minus 3.8 per cent*, felt that limitations on use of the National Park were a disadvantage.

Table B.2 Approximate sampling errors of percentages
(expressed in percentages)

	Ranger married	Non-Ranger married	Ranger single	Non-Ranger single	Total
Reported percentages					
50	6.1	3.8	9.0	13.2	3.8
40 or 60	6.0	3.7	8.8	12.9	3.7
30 or 70	5.6	3.4	8.2	12.2	3.4
20 or 80	4.9	3.0	7.2	10.6	3.0
10 or 90	3.7	2.3	5.4	8.0	2.3
Number of interviews	118	100	64	41	323

Bibliography

Agius, J. (1973) 'Appendix F' in *Report on Workshop on Proposed Regional Centre for the Alligator Rivers Region* Canberra: Cities Commission

Allen, J. B. (1966) *The Company Town in the American West* Norman: University of Oklahoma Press

Altman, J. C. (1983) *Aborigines and Mining Royalties in the Northern Territory* Canberra: Australian Institute of Aboriginal Studies

—— (1985) 'The Impact of Mining Royalties on Aboriginal Economic Development in the Northern Territory' in P. Loveday and D. Wade-Marshall *Economy and People in the North* Darwin: ANU N. Australia Research Unit, pp 61-71

Altman, J. C. and N. Peterson (1984) 'The case for Aboriginal Access to Mining Royalties Under Land Rights Legislation' *Discussion Paper No. 89* Centre of Economic Policy Research, ANU

Anderson, E. M. (1984) 'Environmental Aspects of Regional Resource Developments, A Commonwealth Perspective' in C. C. Kissling et al. (eds) *Regional Impacts of Resource Developments* Sydney: Croom Helm Australia, pp 17-26

Annabell, R. (1971) *The Uranium Hunters* Adelaide: Rigby

Aschmann, H. (1977) 'Views and Concerns Relating to Northern Development' in *Northern Australian Research Bulletin, No. 1* pp 31-57

Atkins, R. (1975) 'Local Government', in R. N. Spann (ed.) *Public Administration in Australia* Sydney: NSW Government Printer

—— (1976) 'Local Government for a New Town: When? What Form? Functions? Why?' in Australian National Commission for UNESCO *New Towns in Isolated Settings* Canberra: AGPS, pp 51-2

—— (1979) *Albany to Zeehan: A New Look at Local Governments* Sydney: Law Book Company

Aungles, S. and I. Szelenyi (1980) 'Industrial Decline and Urban Crisis: Structural Conflicts between the State and Monopoly Capital—The Case of Whyalla' in P. Boreham and G. Dow (eds) *Work and Inequality, Vol. I* Melbourne: Macmillan, pp 86-106

Austin, D. J. (1982) 'A Framework for Australian Studies: Some Reflections on Community, Class and Culture' *Mankind* 13, 3, pp 218-36

—— (1984) *Australian Sociologies* Sydney: Allen and Unwin

Australian Bureau of Statistics (1982) 'Population Count: Jabiru, 21 July 1982' Darwin: ABS Northern Territory Office

179

180   *Yellowcake and crocodiles*

Australian Conservation Foundation (1976) 'Submission by Australian Conservation Foundation to Ranger Uranium Inquiry on Kakadu National Park' Melbourne (mimeo)

Australian Institute of Aboriginal Studies (1980) 'Comments and Recommendations on the Kakadu National Park: Plan of Management' Canberra (mimeo)

—— (1984) *Aborigines and Uranium, Consolidated Report on the Social Impact of Uranium Mining on the Aborigines of the NT* Canberra: AGPS

Australian Labor Party (nd) *Uranium A Fair Trial. Both Sides of the Uranium Debate as Presented in the ALP* Canberra: ALP

Australian Mining Industry Council (1978) 'The Aboriginal Land Rights (Northern Territory) Act. Matters of Serious Concern to the Mining Industry' Canberra: AMIC (mimeo)

—— (1982) 'Aboriginal Land Rights, The Search for a National Consensus' *Mining Review* March

Australian National Commission for UNESCO (1976) *New Towns in Isolated Settings* Canberra: AGPS

Australian National Parks and Wildlife Service (1980) *Kakadu National Park Plan of Management* Canberra: ANPWS

Barker, G. (1978) 'Alligator Rivers Region—Historical Sketch to World War 2' Canberra: Australian Institute of Aboriginal Studies (mimeo)

Barrett, B. (1979) *The Civic Frontier, The Origin of Local Communities and Local Government in Victoria* Melbourne: Melbourne University Press

Barrie, D. R. (1982) *The Heart of Rum Jungle: The History of Rum Jungle and Batchelor in the Northern Territory of Australia* Batchelor, NT: S and D Barrie

Bate, W. (1978) *Lucky City: The First Generation at Ballarat 1851-1901* Melbourne: Melbourne University Press

Bates, G. M. (1983) *Environmental Law in Australia* Sydney: Butterworths

Bath, L. J. (1984) 'Mine Planning and Scheduling at Ranger Uranium Mine—Environmental Requirements and Economics' in *Darwin Conference 1984* Parkville: Australian Institute of Mining and Metallurgy, pp 23-30

Bath, L. J. et al. (1980) 'Development of the Alligator Rivers Uranium Deposits' *Atomic Energy in Australia* 23, 1, pp 12-28

Bauer, F. H. (1964) *Historical Geography of White Settlement in Part of Northern Australia, Part 2, The Katherine—Darwin Region* Canberra: CSIRO

Bell, C. and S. Encel (eds) (1978) *Inside the Whale: Ten Personal Accounts of Social Research* Sydney: Pergamon Press

Bell, C. and H. Newby (1971) *Community Studies* London: Allen and Unwin

Berndt, R. M. (1982) *Aboriginal Sites, Rights and Resource Development* Perth: Academy of the Social Sciences in Australia and University of Western Australia Press

Berndt, R. M. and C. H. Berndt (1954) *Arnhem Land: Its History and Its People* Melbourne: Cheshire

Blainey, G. (1963) *The Rush That Never Ended* Melbourne: Melbourne University Press

Bowman, M. (1976) *Local Government in the Australian States* Canberra: AGPS

—— (ed.) (1981) *Beyond the City: Case Studies in Community Structure and Development* Melbourne: Longman Cheshire

Bradbury, J. H. (1977) 'Instant Towns in British Columbia 1964-1972', PhD thesis, Vancouver: Simon Fraser University

—— (1979) 'Toward an Alternative Theory of Resource-based Town Develop ment in Canada' *Economic Geography* 55, pp 147-66

—— (1980) 'Instant Resource Towns Policy in British Columbia: 1965-1972' *Plan Canada* 20, pp 19-38

—— (1983) 'Housing Policy and Home Ownership in Mining Towns: Quebec, Canada' Department of Geography, McGill University (mimeo)

—— (1984a) 'The Impact of Industrial Cycles in the Mining Sector: the Case of the Quebec—Labrador Region in Canada' *International Journal of Urban and Regional Research* 8, 3, pp 311-30

—— (1984b) 'Declining Single-Industry Communities in Quebec—Labrador, 1979-1983' *Journal of Canadian Studies* 19, 3, pp 125-39

Bradbury, J. H. and I. St Martin (1983) 'Winding Down in a Quebec Mining Town: A Case Study of Schefferville' *Canadian Geographer* XVII, 2 pp 128-44

Brealey, T. B. (1972) *Living in Remote Communities in Tropical Australia: 1. Exploratory Study* Highett, Victoria: CSIRO Division of Building Research

Brealey, T. B. and P. W. Newton (1977) 'The Case for Centralisation' *Mining Review* July, pp 7-9

—— (1978) *Living in Remote Communities in Tropical Australia: The Hedland Study* Melbourne: CSIRO

—— (1981) ' Commuter Mining—An Alternative Way', paper presented to the AMIC Environmental Workshop, Canberra

Brown, J. (1982) 'Infrastructure Policies in the Pilbara' in E. J. Harman and B. W. Head (eds) *State, Capital and Resources* pp 237-55

Bryson, L. (1981) 'Australian Community Studies: A Critique' Urban Research Unit Seminar paper, Canberra: ANU (mimeo)

Bryson, L. and F. Thompson (1976) *An Australian Newtown: Life and Leadership in a Working Class Suburb* Malmsbury: Kibble Books

Bulmer, M. I. A. (1975) 'Sociological Models of the Mining Community' *Sociological Review, New Series* 23, 1, pp 61-92

—— (ed.) (1978) *Mining and Social Change, Durham County in the Twentieth Century* London: Croom Helm

—— (ed.) (1982) *Social Research Ethics* London: Macmillan

Butler, G. J. and L. G. Harris (1984) 'Some Planning Problems in the Provision of Education Facilities in Queensland Resource Development Areas: An Initial Overview' in C. C. Kissling et al. (eds) *Regional Impacts of Resource Developments* pp 69-101

Buttel, F. H. and H. Newby (eds) (1980) *The Rural Sociology of the Advanced Societies: Critical Perspectives* London: Croom Helm

Cameron McNamara and Partners (1972a) *Alligator Rivers Area Regional Centre Feasibility Study, Interim Report* Canberra: Department of National Development

—— (1972b) *Alligator Rivers: A Regional Study to Determine the Feasibility of Establishing a New Town in the Alligator Rivers Region of the Northern Territory* Canberra: Department of National Development

Cameron McNamara-Minenco Joint Venture (1979a) *Housing Design Report* Darwin: Jabiru Town Development Authority
—— (1979b) *Landscaping Programme* Darwin: Jabiru Town Development Authority
—— (1979c) *Town Water Supply* Darwin: Jabiru Town Development Authority
—— (1979d) *Feasiblity Study Lake Development* Darwin: Jabiru Town Development Authority
—— (1979e) *Town Centre Development (Interim Report)* Darwin: Jabiru Town Development Authority
—— (1979f) *Lake Valley Development* Darwin: Jabiru Town Development Authority
Carly, P. J. L. (1977) *'Opening Up': A Report on Considerations Involved in the Progressive Establishment of the Normal Roles of Local Authorities, Government Agencies and the Communities in Mining Towns, Vol I, General; Vol II, Wickham* Perth (mimeo)
Chaloupka, G. (1984) 'Beyond Cultural Sites' in H. Sullivan (ed.) *Visitors to Aboriginal Sites: Access, Control and Management* Canberra: ANPWS, pp 62-73
Chapman, R. J. K. and M. Wood (1984) *Australian Local Government* Sydney: Allen and Unwin
Christian, C. S. and J. M. Aldrick (1977) *Alligator Rivers Study: A Review Report of the Alligator Rivers Region Environmental Fact-Finding Study* Canberra: AGPS
Clark, G. L. (1984) 'A Theory of Local Autonomy' *Annals of the Association of American Geographers* 74, 2, pp 195-208
Clark, T. N. (1974) *Comparative Community Politics* New York: John Wiley
Cole, K. (1975) *A History of Oenpelli* Darwin: Nungalinga Publications
Commonwealth of Australia (1964) *Economic Analysis of Road Development in the Northern Territory Buffalo Area* Canberra: Bureau of Agricultural Economics
—— (1973) *Report on Workshop on Proposed Regional Centre for the Alligator Rivers Region of the Northern Territory, Darwin 18-21 June, 1973* Canberra: Cities Commission
—— (1974) *The Pilbara Study* Canberra: Department of Northern Development
—— (1977) *Uranium Australia's Decision* Canberra: AGPS
—— (1978) *Ranger Uranium Project—Agreement Under Section 44 of the Aboriginal Land Rights (Northern Territory) Act 1976* Canberra: AGPS
—— (1980a) *Representations Received in Connection with the Plan of Management for Kakadu National Park* Canberra: Australian National Parks and Wildlife Service
—— (1980b) *Ranger Uranium Project, Deed to Amend the Government Agreement between the Commonwealth of Australia, Peko-Wallsend Operations Ltd, Electrolytic Zinc Company of Australasia Ltd, Australian Atomic Energy Commission and Energy Resources of Australia Ltd* Canberra: AGPS
—— (1982) *Alligator Rivers Stage II, Land Claim* Canberra: AGPS
Commonwealth Department of Health (1978) *Annual Report 1977-78* Canberra: AGPS

Connell, R. W. (1977) *Ruling Class, Ruling Culture* Cambridge: Cambridge University Press

Coombs, H. C. (1980) 'The Impact of Uranium Mining on the Social Environment of Aboriginals in the Alligator Rivers Region' in S. Harris (ed.) *Social and Environmental Choice* pp 122-35

Courtenay, P. P. (1982) *Northern Australia, Patterns and Problems of Tropical Development in an Advanced Country* Melbourne: Longman Cheshire

Cousins, D. and J. Nieuwenhuysen (1984) *Aboriginals and the Mining Industry: Case Studies of Australian Experience* Sydney: Allen and Unwin

Crommelin, M. and R. D. Nicholson (1981) *Report on Uranium Mining Laws in the Northern Territory* Canberra: Uranium Advisory Council and AGPS

Crough, G. J. and E. L. Wheelwright (1983) 'Australia, Client State of International Capital: A Case Study of the Mineral Industry' in E. L. Wheelwright and K. Buckley (eds) *Essays in the Political Economy of Australian Capitalism* vol. 5, Sydney: ANZ Book Co., pp 15-42

Daly, M. T. (1982) *Sydney Boom, Sydney Bust: The City and Its Property Market 1850-1981* Sydney: Allen and Unwin

Danielson, M. J. (1984) 'Geology and Mineral Potential of the Ranger Project Area, Northern Territory' in *Darwin Conference 1984* Parkville: Australian Institute of Mining and Metallurgy, pp 213-14

Davenport, J. A. and J. Davenport (1979) *Boom Towns and Human Services* Laramie: University of Wyoming

Dearlove, J. (1973) *The Politics of Policy in Local Government* Cambridge: Cambridge University Press

Dennis, N. et al. (1969) *Coal is Our Life: An Analysis of a Yorkshire Mining Community* London: Tavistock

Duncan, S. S. and M. Goodwin (1982) 'The Local State: functionalism, autonomy and class relations in Cockburn and Saunders' *Political Geography Quarterly* 1, 1, pp 77-96

Energy Resources of Australia Ltd (1980) *Prospectus* Sydney: ERA Ltd

—— (1982) *1982 Annual Report* Sydney: ERA Ltd

England, J. L. and S. L. Albrecht (1984) 'Boomtowns and Social Disruption' *Rural Sociology* 49, 2, pp 230-46

Everingham, P. (1982) 'Developing the North' in E. J. Harman and B. W. Head (eds) *State, Capital and Resources* pp 133-48

Feldman, L. D. and M. D. Goldrick (eds) (1972) *Politics and Government of Urban Canada: Selected Readings* 2nd edn, Toronto: Methuen

Forbes, M. A. (1984) 'Visitor Use—Kakadu National Park 1982 and 1983' paper presented to the Kakadu Tourism Seminar, Darwin, September (mimeo)

Forbes, M. A. and P. J. Merrill (1983) 'Interim Report: Visitor Use in Kakadu National Park' Canberra: ANPWS

Formby, J. (1981) 'The Australian Experience' in T. O'Riordan and W. D. R. Sewell (eds) *Project Appraisal and Policy Review* Chichester: Wiley pp 187-225

Fowler, R. J. (1982) *Environmental Impact Assessment, Planning and Pollution Measures in Australia* Canberra: AGPS

Fox, A. M. (1982) 'Kakadu Man and Landscape' Part 1 *Heritage Australia* 1, 2, pp 13-17

—— (1983a) 'Kakadu Man and Landscape' Part 2 *Heritage Australia* 2, 1, pp 23-27

—— (1983b) 'Kakadu Man and Landscape' Part 3 *Heritage Australia* 2, 2, pp 54-57

Fox, R. W. et al. (1976) *Ranger Uranium Environmental Inquiry First Report* Canberra: AGPS

—— (1977) *Ranger Uranium Environmental Inquiry Second Report* Canberra: AGPS

Gale, F. and J. Jacobs (1984) 'Kakadu National Park Visitor Survey Final Report' Canberra: ANPWS

Galligan, B. (1983) 'Utah: A Study of Corporate-State Relations' paper delivered at the Australasian Political Studies Association Conference, Ku-ring-gai College, Sydney

Geertz, C. (1963) *Agricultural Involution* Berkeley: California University Press

Gertler, L. and R. Crowley (1977) *Changing Canadian Cities: The Next 25 Years* Toronto: McClelland and Stewart

Gillett, J. A. and A. D. Robertson (1984) 'The Regional Employment and Population Impacts of Resource Developments' in C. C. Kissling et al. (eds) *Regional Impacts of Resource Developments* pp 53-68

Gilmore, J. S. (1976) 'Boom Towns May Hinder Energy Development: Rural Communities Cannot Handle Sudden Industrialization and Growth Without Help' *Science* 191, pp 535-40

Gilmore, J. S. and M. K. Duff (1975) *Boom Town Growth Management: A Case Study of Rock Springs—Green River, Wyoming* Boulder, Colorado: Westview Press

Godwin, G. W. (1976) 'Statement in Reply to Enquiries Raised as a Result of Evidence Given by Mr Terence Brooks, Assistant Secretary, Urban Development and Town Planning Branch, Dept. of the Northern Territory, 28 May 1976' Evidence submitted to Ranger Uranium Environmental Inquiry

Goodchild Research Studies (1981) 'Local Government in the Northern Territory—Future Directions?' Darwin: Darwin City Council

Government of the Northern Territory (nd) *Draft Proposals on Aboriginals and Land in the Northern Territory* Darwin: Government Printer

—— (nd) *Northern Territory: A Way of Life* Darwin: Government Printer

—— (1981) *Jabiru Town Development Act, No 45/81* Darwin: Government Printer

Government of Western Australia (1962a) *Statutes of Western Australia, Vol 1, 1962, Iron Ore (Mount Goldsworthy) Agreement* pp 43-70

—— (1962b) *Western Australia Parliamentary Debates, Vol 161*

—— (1963a) *Statutes of Western Australia, Vol 1, 1963, Iron Ore (Hamersley Range) Agreement* pp 165-213

—— (1963b) *Western Australia Parliamentary Debates, Vol 165*

—— (1979) *Western Australia Parliamentary Debates, Vol 224*

Graham, S. B. (1978) 'A Comparison of Social Networks in Two Arizona Mining Towns' *Social Science Journal* 15, 2, pp 113-22

—— (1980) 'Community, Conformity and Career: Patterns of Social Interaction in Two Arizona Mining Towns' *Urban Anthropology* 9, 1, pp 1-19

Gribbin, C. C. and J. B. W. Thompson (1980) 'Recruitment of Single Men to

Remote Enterprises: Some Implications for Policy Formulation' *Human Resource Management Australia* 18, pp 45-15

Halligan, J. and C. Paris (eds) (1984) *Australian Urban Politics* Melbourne: Longman Cheshire

Harman, E. J. (1982) 'Ideology and Mineral Development in Western Australia, 1960-1980' in E. J. Harman and B. W. Head (eds) *State, Capital and Resources* pp 167-96

—— (1983) 'The City, State and Resource Development in Western Australia' in P. Williams (ed.) *Social Process and the City* Sydney: Allen and Unwin, pp 114-42

Harman, E. J. and B. W. Head (eds) (1982) *State, Capital and Resources in the North and West of Australia* Perth: University of Western Australia Press

Harris, C. P. (1975) *The Classification of Australian Local Authorities* Canberra: Centre for Research on Federal Financial Relations, ANU

—— (1978) *Local Government and Regionalism in Queensland 1859-1977* Canberra: Centre for Research on Federal Financial Relations, ANU

Harris, S. (1972) *This is Our Land* Canberra: ANU Press

Harris, S. (ed.) (1980) 'Social and Environmental Choice: The Impact of Uranium Mining in the Northern Territory' *CRES Monograph 3* Centre for Resource and Environmental Studies, ANU

Hayward, N. L. (1984) 'Staffing Arrangements at Queensland Mines Limited's Nabarlek Uranium Treatment Plant' *Darwin Conference 1984* Parkville: Australian Institute of Mining and Metallurgy, pp 9-16

A. A. Heath and Partners (nd) *Jabiru New Town, Advanced Design Study Interim Report* Brisbane: Cameron McNamara and Partners

—— (1978) *Advanced Design Study Final Report* Brisbane: Cameron McNamara and Partners

A. A. Heath and Partners and Willis, Heathwood and Partners (1973) *Design Study: Regional Town Project, Alligator Rivers, NT* Canberra: Cities Commission

Heatley, A. (1979) *The Government of the Northern Territory* Brisbane: University of Queensland Press

—— (1982) 'Parties and Development in the Northern Territory' in E. J. Harman and B. W. Head (eds) *State, Capital and Resources,* pp 197-214

Heppell, M. (ed.) (1979) *A Black Reality, Aboriginal Camps and Housing in Remote Australia* Canberra: Australian Institute of Aboriginal Studies

Howitt, R. and J. Douglas (eds) (1983) *Aborigines and Mining Companies in Northern Australia* Sydney: Alternative Publishing Cooperative

Idris-Soven, A. (1978) *The World as a Company Town: Multinational Corporations and Social Change* The Hague: Mouton

Jabiru Town Development Authority (1980) *First Annual Report 1979* Darwin: JTDA

—— (1981a) *Annual Report 1979-80* Darwin: JTDA

—— (1981b) *Third Annual Report 1980/81* Darwin: JTDA

—— (1982) *Fourth Annual Report* Darwin: JTDA

—— (1983) *Fifth Annual Report* Darwin: JTDA

—— (1985) *Annual Report 1983-84* Darwin: JTDA

Jackson, R. T. (1984) 'Ok Tedi: Lessons Hardly Learnt' in C. C. Kissling et al. (eds) *Impacts of Resource Developments,* pp 117-33

Jones, M. A. (1977) *Organizational and Social Planning in Australian Local Government* Melbourne: Heinemann

Jones, R. (ed.) (1980) *Northern Australia: Options and Implications* Canberra: Research School of Pacific Studies, ANU

—— (1984) 'From Xanadu to Kakadu: Interpretative Ideas for the Prehistory of a Pleasure Park' in H. Sullivan (ed.) *Visitors to Aboriginal Sites* pp 94-113

Keen, I. (1980) 'The Alligator Rivers Aborigines—Retrospect and Prospect' in R. Jones (ed.) *Northern Australia* pp 171-86

Kennedy, B. E. (1978) *Silver, Sin and Sixpenny Ale, A Social History of Broken Hill, 1883-1921* Melbourne: Melbourne University Press

—— (1981) 'How Distance Shaped a Community: Broken Hill's First Ten Years' in M. Bowman (ed.) *Beyond the City* pp 114-26

—— (1984) *A Tale of Two Mining Cities* Melbourne: Melbourne University Press

Kerr, C. and A. Siegel (1954) 'The Inter-Industry Propensity to Strike: An International Comparison' in A. Kornhauser et al. (eds) *Industrial Conflict* New York: McGraw Hill, pp 189-212

Keesing, N. (ed.) (1971) *History of Australian Gold Rushes by Those Who Were There* Sydney: Angus and Robertson

Kilmartin, L. and D. Thorns (1978) *Cities Unlimited: The Sociology of Urban Development in Australia and New Zealand* Sydney: Allen and Unwin

Kriegler, R. J. (1980) *Working for the Company, Work and Control in the Whyalla Shipyard* Melbourne: Oxford University Press

Langrod, G. (1953) 'Local Government and Democracy' *Public Administation* XXXI, pp 25-33

Lansing, J. B., R. W. Marans and R. B. Zehner (1970) *Planned Residential Environments* Ann Arbor: Institute for Social Research, University of Michigan

Layman, L. (1982) 'Changing Resource Development Policy in Western Australia, 1930s-1960s' in E. J. Harman and B. W. Head (eds) *State, Capital and Resources* pp 149-65

Lea, J. P. (1979) 'Self-help and Autonomy in Housing: Theoretical Critics and Empirical Investigators' in H. S. Murison and J. P. Lea (eds) *Housing in Third World Countries: Perspectives on Policy and Practice* London: Macmillan, pp 49-53

—— (1982) 'Government Dispensation, Capitalist Imperative or Liberal Philanthropy? Responses to the Black Housing Crisis in South Africa' in D. M. Smith (ed.) *Living Under Apartheid: Aspects of Urbanisation and Social Change in South Africa* London: Allen and Unwin, pp 198-216

—— (1983) 'Customary Tenure and Urban Housing Land: Partnership and Participation in Developing Societies' in S. Angel et al. (eds) *Land for Housing the Poor* Singapore: Select Books, pp 54-72

—— (1984) 'Housing Needs and Social Demands in the Special Development Projects of the Northern Territory' in D. W. Drakakis-Smith (ed.) *Housing in the North: Policies and Markets* Darwin: N. Australia Research Unit, ANU, pp 125-57

Lea, J. P. and R. B. Zehner (1983) 'The Planning of Resource-based Communities in Australia: Identifying the Issues' *Occasional Paper No. 12* Planning Research Centre, University of Sydney

—— (1984) 'Submission under Sub-Section II (2) of the National Parks and Wildlife Conservation Act 1975 in respect of the Plan of Management for Kakadu National Park' Canberra: ANPWS

—— (1985) 'Democracy and Planning in a Small Mining Town: the Governance Transition in Jabiru, NT' in P. Loveday and D. Wade-Marshall (eds) *Economy and People in the North* Darwin: N. Australia Research Unit, ANU, pp 222-47

Leichhardt, L. (1847) *Journal of an Overland Expedition in Australia from Moreton Bay to Port Essington* London: T. W. Boone

Levitus, R. (1982) 'Everybody Bin All Day Work' Canberra: Australian Institute of Aboriginal Studies

Lloyd, P. J. (1984) *Mineral Economics in Australia* Sydney: Allen and Unwin

Loveday, P. (1982) *Promoting Industry* St Lucia: Queensland University Press

Loveday, P. and J. P. Lea (1985) *Aboriginal Housing Needs in Katherine* Darwin: N. Australia Research Unit, ANU

Lucas, R. A. (1971) *Minetown, Milltown, Railtown: Life in Canadian Communities of Single Industry* Toronto: University of Toronto Press

Macknight, C. C. (1976) *The Voyage to Marege* Melbourne: Melbourne University Press

McIntyre, A. J. and J. J. McIntyre (1944) *Country Towns of Victoria* Melbourne: Melbourne University Press

McPhail, I. (1978) 'Local Government' in P. N. Troy (ed.) *Federal Power in Australian Cities* Sydney: Hale and Iremonger, pp 105-16

—— (nd) 'Revision of the Northern Territory Local Government Ordinance', (mimeo)

Markusen, A. R. (1980) 'The Political Economy of Rural Development: The Case of Western US Boomtowns' in F. H. Buttel and H. Newby (eds) *Rural Sociology of the Advanced Societies* pp 405-30

Matthews, R. (ed.) *Local Government in Transition: Responsibilities, Finances, Management* Canberra: Centre for Research on Federal Financial Relations, ANU

Mercer, J. (1982) 'The Community Needs of Remote Mining Towns' paper delivered at the 'Coal and Society' Seminar, University of Sydney, November 23-25

Millard, S. (1981) 'Public Participation in Planned Community Development' in M. Bowman (ed.) *Beyond the City* pp 188-201

Montague, M. (1977) 'Barcaldine: Internal Migration and an Australian Rural Community' PhD thesis, St Lucia: University of Queensland

Mount Newman Mining Co. Ltd (1981) *Employees Home Ownership Scheme* Port Hedland: Mount Newman Mining Co. Ltd

MSJ Keys Young Planners (1978) 'Mount Newman Mining Pty Ltd, Home Ownership Program' Sydney: MSJ Keys Young

Murray, M. D. et al. (1980) 'Living in the Pilbara: The Employees View' Port Hedland: Mount Newman Mining Co. Ltd

Neil, C. C. (1982) 'Housing Symbolism in New Remote Mining Communities in Australia: Implications for Innovative Versus Conventional Design and Siting of Houses in Harsh Environments' *Journal of Environmental Psychology* 2, pp 201-20

Neil, C. C. and T. B. Brealey (1982) 'Home Ownership in New Resource

Towns: Will it Change or Reinforce Existing Social Trends?' *Human Resource Management Australia* February, pp 38-44

Neil, C. C. et al. (1982) 'The Development of Single Enterprise Resource Towns' *Occasional Paper No. 25* Highett, Victoria: CSIRO Division of Building Research

Neil, C. C. et al. (1984) 'Population Stability in Northern Australian Resource Towns: Endogenous Versus Exogenous Influences?' in D. Parkes (ed.) *Northern Australia: The Arenas of Life and Ecosystems on Half a Continent* Sydney: Academic Press, pp 363-93

Neilson, L. (1983) 'How Local Government Comes to Company Towns' *Australian Planner* August/September, pp 91-92

Newton, P. W. (1977) 'Housing Costs in Remote Areas of Australia' submission to Committee of Inquiry into Housing Costs, Canberra (mimeo)

Newton, P. W. and T. B. Brealey (1979) 'Commuter Mining: Toward the Rational Development of Resource Regions in Tropical Australia' *Mining Review* April, pp 6-11

Noranda Australia (1978) *Koongarra Project, Draft Environmental Impact Statement* Melbourne: Noranda Australia

Northern Land Council (nd) 'Land Rights Wrongs', Darwin: special edition of *Land Rights News*

Northern Territory Electricity Commission (1979) 'Report on Power Development for the Alligator Rivers Uranium Area' Perth: Merz and Mclellan and Partners

Northern Territory Liquor Commission (1983) 'Report of Consultations Pursuant to s. 31. 3. 13 of Kakadu National Park Plan of Management Held at Jabiru on 12 and 28 January 1983' Darwin: NT Liquor Commission

Northern Territory News (1979) 'Jabiru, A Town in the Making' Special *NT News* Feature, 15 December

—— (1981) 'Jabiru, One Year Later' Special *NT News* Feature, 21 April

Oeser, O. A. (1981) 'Single Industry Company Towns: the Psychology of Community Development' in M. Bowman (ed.) *Beyond the City* pp 203-17

Oeser, O. A. and F. E. Emery (eds) (1954) *Social Structure and Personality in a Rural Community* London: Routledge and Kegan Paul

O'Faircheallaigh, C. (1985) 'Managing the Impact of Large Mines in Remote Areas: Case Studies for Papua New Guinea and the Northern Territory' paper presented to the International Regional Management Conference of the Australian Institute of Management, Darwin, 3-6 September

Ollier, C. (1980) 'Environmental Problems of the Mining Industry, Especially in Northern Australia' in R. Jones (ed.) *Northern Australia: Options and Implications* pp 161-70

Ovington, J. D. and R. W. Boden (1979) 'Alligator Rivers Region in Northern Australia—A Case Study in Land Use Planning, *Landscape Planning* 6, pp 299-312

Oxley, H. G. (1974) *Mateship in Local Organisation* St Lucia: University of Queensland Press

Paget, G. and R. A. Rabnett (1983) 'The Need for Changing Models of Planning: Developing Resource Based Communities' *UBC Planning Papers, Canadian Planning Issues, No. 6,* Vancouver

Paget, G. and B. Walisser (1984) 'The Development of Mining Communities in British Columbia: Resilience Through Local Governance' in *Mining Communities: Hard Lessons for the Future* Proceedings of the Twelfth CRS Policy Discussion Seminar, Kingston, Ontario, 27-29 September 1983, pp 96-150

Palmer, K. (ed.) (1985) 'Aborigines and Tourism: A Study of the Impact of Tourism on Aborigines in the Kakadu Region, Northern Territory' Darwin: Northern Land Council

Paris, C. (1983) 'Urban Politics: Where to Now?' in P. Williams (ed.) *Social Process and the City* pp 216-25

Parkin, A. (1980) 'Who Governs Australia's Cities?' in A. Parkin et al. (eds) *Government, Politics and Power in Australia* Melbourne: Longman Cheshire, pp 374-89

Parsons, D. (1978) 'Inside the Ranger Negotiations' *Arena* 51, pp 134-43

Peace, A. (1985) 'Small Town Politics and Crises of Legitimation' *Journal of Australian Studies* 16, pp 84-96

Peat Marwick Mitchell Services (1982) 'Alligator Rivers Tourist Accommodation Study' Darwin: NT Development Corporation

Porteous, J. D. (1970) 'The Nature of the Company Town' *Transactions of the Institute of British Geographers* 51, pp 127-42

—— (1974) 'Social Class in Atacama Company Towns' *Annals of the Association of American Geographers* 64, 3, pp 409-17

Powell, A. (1982) *A Far Country* Melbourne: Melbourne University Press

Power, J., R. Wettenhall and J. Halligan (eds) (1981) *Local Government Systems of Australia* Canberra: AGPS

Ranger Uranium Mines Ltd (1975) *Environmental Impact Statement February 1974* Sydney: Ranger Uranium Mines

Reese, M. H. and J. C. Cummings (1979) 'Energy Impacted Housing' in J. A. Davenport and J. Davenport (eds) *Boom Towns and Human Services* pp 63-78

Rhodes, R. A. W. (1980) 'Some Myths in Central-Local Relations' *Town Planning Review* 51, pp 270-85

Richmond, W. H. and P. C. Sharma (eds) (1983) *Mining in Australia* St Lucia: Queensland University Press

Riffel, J. A. (1975) *Quality of Life in Resource Towns* Ottawa: Information Canada

Roberts, J. (1981) *Massacres to Mining: The Colonisation of Aboriginal Australia* Blackburn, Victoria: Dove Communications

Roberts, S. (1984) 'Statement to the Ranger Uranium Environmental Inquiry' in P. J. Lloyd (ed.) *Mineral Economics in Australia* pp 136-7

Robinson, I. (1984) 'Planning Strategies for Town and Regional Development in Resource Regions' Seminar Paper presented 24 May, CSIRO Division of Building Research, Highett, Victoria

Rogers, P. H. (1973) *The Industrialists and the Aborigines* Sydney: Angus and Robertson

Roxby Management Services Ltd (1982) *Olympic Dam Project, Draft Environmental Impact Statement* Adelaide: Kinhill-Stearns Roger Joint Venture

Ryan, G. R. (1972) 'Ranger 1: A Case History' in *Uranium Prospecting*

Handbook London: Institute of Mining and Metallurgy, pp 296-300

Ryan, B. (1980) 'Government Intervention in Rural Australia' in W. P. Avery et al. (eds) *Rural Change and Public Policy* New York: Pergamon Press, pp 36-61

Saddler, H. (1978) 'Public Participation in Technology Assessment with Particular Reference to Public Inquiries' *General Paper R/GP9* Canberra: ANU Centre for Resource and Environmental Studies

—— (1980) 'Implications of the Battle for the Alligator Rivers: Land Use Planning and Environmental Protection' in R. Jones (ed.) *Northern Australia: Options and Implications* pp 187-200

Saddler, H. and J. P. Kelly (1983) 'The Uranium Decision' in W. H. Richmond and P. C. Sharma (eds) *Mining and Australia* pp 257-83

Salaman, G. (1971) 'Two Occupational Communities: Examples of a Remarkable Consequence of Work and Non-Work' *Sociological Review* 19, pp 389-407

Sandercock, L. (1974) 'Politics, Planning and Participation' *Australian Quarterly* 46, 3, pp 48-64

Saunders, P. (1983) 'On the Shoulders of Which Giant? The Case for Weberian Political Analysis' in P. Williams (ed.) *Social Process and the City* pp 41-63

—— (1984) 'The Crisis of Local Government in Melbourne: the Sacking of the City Council' in J. Halligan and C. Paris (eds) *Australian Urban Politics* pp 88-109

Serle, G. (1963) *The Golden Age, A History of the Colony of Victoria, 1851-1861* Melbourne: Melbourne University Press

—— (1971) *The Rush to be Rich, A History of the Colony of Victoria, 1883-1889* Melbourne: Melbourne University Press

Simpson, E. J. (1980) 'Jabiru Town Planning' in S. Harris (ed.) *Social and Environmental Choice* pp 87-107

Smith, T. (1979) 'Forming a Uranium Policy: Why the Controversy?' *Australian Quarterly* 51, 4, pp 32-50

Soovere, K. J. (1967) 'Community Power in Mount Isa' PhD thesis, St Lucia: University of Queensland

Spencer, W. B. (1928) *Wanderings in Wild Australia* London: Macmillan

Stacpoole, H. J. (1971) *Gold at Ballarat, The Ballarat East Goldfield Its Discovery and Development* Melbourne: Lowden Publishing

Stanley, O. (1982) 'Royalty Payments and the Gagudju Association' in P. Loveday (ed.) *Service Delivery to Remote Communities* Darwin: N. Australia Research Unit, ANU, pp 36-49

Stimson, R. J. (1982) *The Australian City: A Welfare Geography* Melbourne: Longman Cheshire

Stockbridge, M. et al. (1976) *Dominance of Giants, A Shire of Roebourne Study* Perth: Department of Social Work, University of Western Australia

Stolz, G. E. (1983) 'Jabiru-Mining Town with a Difference' *The Australian*, 1 July

Sullivan, H. (ed.) (1984) *Visitors to Aboriginal Sites: Access, Control and Management. Proceedings of the 1983 Kakadu Workshop* Canberra: ANPWS

Supervising Scientist for the Alligator Rivers Region (1982) *Office of the Supervising Scientist. A Report to the Prime Minister by the Australian Science and Technology Council* Canberra: AGPS

Tatz, C. (1982) *Aborigines, Uranium and Other Essays* Melbourne: Heinemann

Tatz, C. et al. (1982) 'Oral Evidence to House of Representatives Standing Committee on Aboriginal Affairs (Reference: Fringe-Dwelling Aboriginal Communities)' Canberra, 25 February

Thompson, H. M. (1981) '"Normalisation": Industrial Relations and Community Control in the Pilbara' *Australian Quarterly* 53, 3, pp 301-24

—— (1983) 'The Pyramid of Power: Transnational Corporations in the Pilbara' in E. L. Wheelwright and K. Buckley (eds) *Essays on the Political Economy of Australian Capitalism, Vol. 5* pp 75-100

Trollope, A. (1873) *Australia and New Zealand* Melbourne: George Robertson

Underhill, J. A. (1976) *Soviet New Towns: Housing and National Urban Growth Policy* Washington DC: US Dept of Housing and Urban Development

Urlich Cloher, D. (1979) 'Urban Settlement in Lands of "Recent Settlement" — An Australian Example' *Journal of Historical Geography* 5, 3, pp 297-314

Veno, A. and N. F. Dufty (1982) 'Mine Towns in Australia's North-West: Options for Operation and Quality of Life' paper presented to 52nd ANZAAS Congress, Macquarie University, Sydney

Von Sturmer, J. R. (1982) 'Aborigines in the Uranium Industry: Toward Self-Management in the Alligator River Region?' in R. M. Berndt (ed.) *Aboriginal Sites, Rights and Resource Development* pp. 69-116

Walker, A. (1945) *A Social Survey of Cessnock* Melbourne: Melbourne University Press

Warburton, C. (1934) *Buffaloes: Adventure and Discovery in Arnhem Land* Sydney: Angus and Robertson

Warren, R. L. (1956) 'Toward a Typology of Extra-Community Controls Limiting Local Community Autonomy' *Social Forces* XXIV, 4, pp. 338-41

—— (1972) *The Community in America* Chicago: Rand McNally

Weaver, S. M. (1984) 'Progress Report: the Role of Aboriginals in the Management of Coburg and Kakadu National Parks, Northern Territory, Australia' seminar paper delivered at the N. Australia Research Unit, Darwin, 30 July

Weber, M. (1949) *The Methodology of the Social Sciences* New York: The Free Press

Western, J. S. (1973) 'What White Australians Think' in G. E. Kearney et al. (eds) *The Psychology of Australian Aborigines* Sydney: Wiley

—— (1983) *Social Inequality in Australian Society* Melbourne: Macmillan

White C. (1967) 'The Prehistory of the Kakadu People' *Mankind* 6, 9, pp 426-31

Wild, R. A. (1974) *Bradstow: A Study of Status, Class and Power in a Small Australian Town* Sydney: Angus and Robertson

—— (1981) *Australian Community Studies and Beyond* Sydney: Allen and Unwin

—— (1983) *Heathcote: A Study of Local Government and Resident Action in a Small Australian Town* Sydney: Allen and Unwin

Williams, C. (1979) 'Capitalism, Patriarchy and the Working Class: A Sociological Study of Open Cut Coal Mining in Queensland' PhD thesis, St Lucia: University of Queensland

—— (1981) *Open Cut: the Working Class in an Australian Mining Town* Sydney: Allen and Unwin

Williams, P. (ed.) (1983) *Social Process and the City* Sydney: Allen and Unwin

Wood, M. (1979) 'Local Government in Western Australia' in R. Pervan and C. Sharman (eds) *Essays on Western Australian Politics* Perth: University of Western Australia Press, pp 97-125

Woods, D. T. (1979) 'The Ranger Uranium Project' *Australian Institute of Mining and Metallurgy Bulletin* 432, pp 11-15

Woods, D. T. et al. (1978) 'The Ranger Uranium Environmental Inquiry and the Environment Protection (Impact of Proposals) Act 1974' in *Environmental Engineering Conference 1978—Environmental Enquiry* Sydney: Institution of Engineers, pp 25-30

Woodward, A. E. (1973) *Aboriginal Land Rights Commission—First Report* Canberra: AGPS

—— (1974) *Aboriginal Land Rights Commission—Second Report* Canberra: AGPS

Young, E. A. and E. K. Fisk (eds) (1982a) *Town Populations, The Aboriginal Component in The Australian Economy, No. 2* Canberra: Development Studies Centre, ANU

—— (1982b) *Small Rural Communities, The Aboriginal Component in the Australian Economy, No. 3* Canberra: Development Studies Centre, ANU

Young, R. I. (1983) 'The Ranger Project, A Case Study' in *Project Development Symposium* Sydney: Australian Institute of Mining and Metallurgy Sydney Branch, pp 261-69

Zehner, R. B. (1971) 'Neighborhood and Community Satisfaction in New Towns and Less Planned Suburbs' *Journal of the American Institute of Planners* 37, 6, pp 379-85

—— (1977) *Indicators of the Quality of Life in New Communities* Cambridge: Ballinger

Zehner, R. B. and F. S. Chapin Jr (1974) *Across the City Line: A White Community in Transition* Lexington: Lexington Books

Zehner, R. B. and J. P. Lea (1983) 'The Planning and Administration of Mining Communities' Report prepared for Australian Mineral Industries Research Association, School of Town Planning University of NSW

—— (1984) 'Jabiru Community Survey' School of Town Planning, University of NSW

Zillman, D. N. and S. Solomon (1983) 'The Impact on Communities of Major Developments' paper delivered to 53rd ANZAAS Congress, Perth

Index

Aborigines, alcohol and, 74, 87–8, 129, 146–9; housing of, 71–3, 149; in Jabiru, 60–1, 71–4, 92, 104, 106, 129, 135–6, 146–50, 156–8, 161–4, 169–70, 175; land rights and, xv, xvii, 10, 33, 47, 50–1, 55, 91–2, 143–4, 146, 149, 156, 164, 167; population, xv, 56
Aboriginal Land Rights Commission, NT, 37, 47
Aboriginal Land Rights (Northern Territory) Act, 1976, xvii, 37, 50–1, 55, 56
Adelaide, South Australia, 110
Agius, J., 46, 112
alcohol, Aborigines and *see* Aborigines; abuse of, 83, 129, 148; in Jabiru, 87, 89, 130, 137–8, 145–6, 151
Aldrick, J.M., 171
Allen, G., xii
Alligator Rivers Region, Aboriginal presence in, 33, 41, 56; Co-ordinating Committee for, xvii, 88; history of, 34–6; literature on, xxii–xxiii, 171–2; location of, 34–5; regional centre in, 43–7; uranium discoveries in, xv, 36–42
Alligator Rivers Region Research Institute, 88
Alligator Rivers Stage I Land Claim, 50, 56
Altman, J.C., xxi, xxiii, 50, 55–6, 92, 171
Alyangula, NT, 60, 167
Annabell, R., 36, 171
anomalies, radiometric, 36–9, 41–2, 45
Anthony, Hon. D., 79
anti-nuclear movement, xxi
Arnhem Highway, 36, 42–3, 58, 66, 71, 140, 166

Arnhem Land, xv, 36, 52, 55, 140, 167
Arthur, W.S., 61
asbestos, 21
Atkins, R., 4, 24
Atkinson, D.N., 79, 83
Aungles, S., 27
Austin, D.J., xix, xx, 3
Australian Atomic Energy Commission, 47, 49, 88
Australian Bureau of Statistics, 103, 107
Australian Conservation Foundation, 49, 88
Australian Institute of Aboriginal Studies, xxiii, 56, 88, 171
Australian Mineral Industries Research Association, xviii
Australian Mining Industry Council, 55, 88
Australian National Commission for UNESCO, 135
Australian National Parks and Wildlife Service, xii, xvii, xxiii, 53, 58, 80, 85, 87–9, 90, 141, 145, 149, 157, 160, 164, 166, 171, 175
Australian Research Grants Scheme, xii, xix
Australian Society of Engineers, 98, 134
Australian Uranium Producers Forum, 49
Australian urban studies, 7, 163
autonomy, xiv, 9, 14, 18, 87, 159

Baily, P., xii
Balmanidbal, P., 41–2
Ballarat, Victoria, 15, 17, 163
Baralil Creek, 58, 66
Barker, G., 171
Barrett, B., 13, 15–17, 28–9
Barrie, D.R., 6
Batchelor, NT, 6, 69

Bate, W., 15
Bath, L.J., 172
Bauer, F.H., 171
bauxite mining, 9, 78, 92
Bell, C., xix
Bendigo, Victoria *see* Sandhurst
Berndt, C.H., 171
Berndt, R.M., 171
Blain, A.D., 61
Blainey, G., 13, 15
Boden, R.W., 171
Border Store, East Alligator, 49, 52
Bougainville, 102
Boulder-Eyre, member for, 20
Bowen Basin, 6, 43, 111, 119
Bowman, M., xiv
Bradbury, J.H., 4, 8, 25, 159
Brealey, T.B., 6, 119, 127, 135, 173
Bridge Autos Jabiru, 85
Brink, M., xii
Brisbane, Queensland, 110
British Columbia, 163-4
Broken Hill, NSW, 17-18
Broken Hill Propriety Company Ltd,
    17, 27, 49
Brooks, T., 47
Bryson, L., 6
buffalo, 34, 36, 49
Bulmer, M.I.A., 3-4, 10-11, 97, 165-9
Bureau of the Northern Land Council
    see Northern Land Council

Cahill, Paddy, 34, 71
Cameron McNamara and Partners, 48,
    56, 58, 172
Cameron McNamara-Minenco Joint
    Venture, 56, 61, 69-70, 172
Canada, 4-5, 25, 163-5
Canberra, ACT, 7, 9, 64, 86, 118, 131,
    136, 144, 170
capital, accumulation, 29; industrial, 25;
    mining, 14-15, 28, 66, 90; multi-
    national, 7, 18-19, 28, 159
capitalist society, 8, 11, 28
caravans, 113, 116, 118, 174, 176
Carly, P.J.L., 21-2
Carly Report the, 21-6
Castlemaine, Victoria, 13, 15-17, 163
cattle industry, 34
Centennial Park, NSW, 130
Cessnock, NSW, 6
Chaloupka, G., xxiii
Chapin, F.S. Jnr, xx, 132

Chapman, R.J.K., 8, 29
Chief Minister, Northern Territory *see*
    Everingham P.A.E.
Charleville, Queensland, 111
Choice magazine, 136
Christian, C.S., 171
Cities Commission, 37, 46, 53
class, xix, 8, 11, 28, 137-8, 159
closed towns, xv, 8, 10, 20-1, 23, 34,
    51, 59, 89, 92, 109, 111, 162
Cole, K., 71, 171
collectivism, 10, 91
Collins, R., 59
Commonwealth, Department of
    Aboriginal Affairs, 88; Health, 58;
    Home Affairs and Environment, 88;
    Trade and Resources, 88
Commonwealth of Australia, xx, 36, 42,
    70, 171-2
community, xiv, xviii, 3, 11, 28, 97, 99,
    125, 130-5, 138, 151-3, 158, 163,
    165, 174; spirit, 109-11, 132, 137,
    150-3; studies, xix, 3, 4-7; survey *see*
    Jabiru Community Survey
Connell, R.W., 8
conservation *see* environment
Conservation Council of South
    Australia, 49
Cooinda, 51-2, 161
Coombs, H.C., 171
corporatist interests, 9, 87, 90-1
cost sharing agreement, xviii, 62, 64, 74,
    160
Country Liberal Party, NT, 78, 81, 87,
    162
Court, Sir Charles, 19-20
Court, G.A., xii, 79
Cousins, D., xxi
Croker Island, NT, 55
Crown land, xvii, 24, 28, 47, 164
CSR Ltd, 21
Cummins, J., 41-2
Cyclone Tracy, 91

Dadbe, xxii-xxiii, 41
Daly, M.T., 18
Dampier, WA, 7, 22
Dampier Mining Company Ltd, 49
Danielson, K.T., xii, 79
Danielson, M.J., 38, 172
Darwin, NT, 7-9, 36, 42, 46, 55, 61,
    69-70, 78, 81, 83, 90-1, 110, 127,
    133, 136, 138-40, 161-3, 166

Darwin Conservation Society, 43
decentralisation, 91, 161
democracy, xvii, 15, 17, 24–5, 51, 90, 159, 163
Denison Australia, xv, xviii, 65, 88, 90, 160
Dennis, N. et al., 4
Department of the Northern Territory, 46–7, 59
Director of the Australian National Parks and Wildlife Service, xvii, 51, 56, 58–9, 79, 87–9, 141
diversification, 5, 8, 21, 23, 50–1, 54, 90–1, 161–2, 164, 166, 170
Djidbidjidbi, xxiii, 41
Douglas, J., xxiii, 55, 171
Dyer, A.J., 71

East Alligator River, 34–5, 140
East Pilbara Shire, WA, 22, 24–7
El Sherana, 35–6
elections, Hill End, NSW, 170; Jabiru Town Advisory Council, 1982, xix, 79, 82, 151–2, 165; Jabiru Town Council, 1984, xix, 53, 78–9, 84, 149, 151–2, 165; Nhulunbuy, 1980, 78; Victorian goldtowns, 16–17
Electrolytic Zinc Co. of Australasia Ltd, 37, 61
Encel, S., xix
encephalitis, 58
Energy Resources of Australia Ltd, xviii, 56, 79, 80, 82–4, 88, 97–9, 134, 160–1
environment, conservation of, 43, 46, 160, 164, 170; environmentalism, xxiii, 33, 170; environmental standards, 41, 58; impact on, xxii, 41, 43, 47, 50–1, 164
Environment Protection (Impact of Proposals) Act, 1974, xxiii, 37, 47
ethnography, 3
Eureka Stockade, 15
European mining towns, 10–11, 97, 159, 166–8
Everingham, P.A.E., xix, 8, 51, 53–4, 59, 64, 79–84, 86, 89, 91, 161–2
EZ Industries Ltd, 56

family, 11, 70, 101, 107, 119, 121, 123–5, 139, 167–8, 174–5
feminist concepts, 6
Finger, M., 59

Fisk, B., xii
fly-in/fly-out, 90
Forbes, M.A., xxiii, 171
Formby, J., 50, 171
Fort Dundas, NT, 34
Fort Wellington, NT, 34
Fox, A.M., 171
Fox, R.W. et al., xv, 33, 47, 49, 165, 171
Fox Inquiry *see* Ranger Uranium Environmental Inquiry
Franklin Dam, Tasmania, xxii
Fraser Island, Queensland, xxiii
Freestone, R., xii
Friends of the Earth, 49

Gagudju Association, xviii, xxiii, 56, 71, 73, 81, 88, 90, 92, 126, 149, 158, 162, 175
Gale, F., xxiii, 171
Galligan, B., 10, 87
Geertz, C., xxii
geiger counter, 35
geography, xx, 3, 7
Geopeko Ltd, 36–8
Georgetown exploration camp, 37–8
Geraldton, WA, 23
Getty Oil, 37
Gillespie, D., xii
Godwin, G.W., 53
gold rush, Victoria, 13, 15
Goldsworthy Agreement, 20
Goldsworthy Mining Company Ltd, 19, 111
Gondwana Joint Venture, 37
Governance, Canadian examples of, 163–5; community participation in, xiv, xviii–xix, 47, 54, 150–3, 165; development authority model of, 7, 21, 46, 88, 150, 159–61; transfer of functions, Jabiru, xvii, 7, 10, 78, 80–1, 161–3, 164, 174
Government of Western Australia, 19–20, 24
Gove Peninsula, NT, 9, 46, 78, 162
Gray, M., 79
Green, D.B., xii, 79
Groote Eylandt Mining Company Ltd, 49
Gunpowder, Queensland, 108, 110
gunwingga, 71

Hamersley Iron Ltd, 22, 111

Hamersley Range Agreement, 20
Harman, E.J., 18–19
Harris, S., 171
Hawke, R.J.L., 54, 84, 87, 89
Heath, A.A. and Partners, 43, 46, 56, 58, 61, 112, 172
Heatley, A., 8, 161
Hebblewhite, R., 81
Hill, Mr and Mrs K. *see* Border Store, East Alligator
Hill End, NSW, 170
Hinton, D., xii
housing, Aboriginal *see* Aborigines; airconditioning in, 60, 69–70, 83, 112, 117, 119, 125; company, 20, 23, 60, 82, 98–9, 139, 168; costs of, 26, 28, 65–6, 70; demountables, 113, 115–7; design of, 66, 68–70, 110, 112–9, 149; evaluation of, 97, 111–9; Jabiru in, 56, 60, 64, 66–70, 89–90, 98–9, 110–9, 139, 168–9, 174; Mt Newman ownership scheme, 23–6, 111; privacy in, 116–8; private sector, 89, 111; public, 60, 90, 112, 139, 168; rents, 83, 98–9, 111, 119; single person for, 112–5, 121–5, 174, 176; social mix in, 60, 65, 67, 112, 122–3
Houston, W., xii
Howieson, M., 24–5
Howitt, R., xxiii, 55, 171
Hull, W.J.F., 60

ideal types, 9, 10, 97, 165–9
ideology, 9–10, 91–2
individualism, 10, 91
institutional analysis, xvii, xx, 87–8, 159–61, 163
interviews, xii–xiii, xix, 7, 99, 106, 110, 135–6, 138, 140, 146, 156, 173–8
inter/intragovernmental issues, xvii, 9, 18, 55, 74, 86–92, 159
iron ore, 13, 18–19, 27, 29
Iron Ore (Mount Newman) Agreement Amendment Act, 1979, 23, 25
isolation *see* remote communities

Jabiluka, xv, 37, 42, 84, 86
Jabiru, Aborigines in *see* Aborigines; by laws, 74, 81; construction of, xxiv, 55–60, 164; clubs in, 126–7, 129–30, 146, 149, 151–3, 167; dogs in, 137, 139, 141, 156; education in, 104, 106, 108, 135, 139, 156; employment in, 104, 106–7, 139, 148, 150, 166–7;

facilities, satisfaction with, 104, 106–7, 125–34, 148, 150, 156, 166–7; financing of, 61–5, 160–1; future for, 157, 169–70; government, participation in, xviii–xix, 47, 74, 79, 81, 151–3, 156–7, 165; headworks in, 61, 64, 66, 78, 80, 85, 89, 160; head lease, 74, 78–9, 160; health in, 58, 81, 83, 126–7, 137–8, 156; hotel/motel in, 92, 146, 162; housing in *see* housing; incomes in, 105, 107, 168–9; Kakadu National Park, controls and, 53, 92, 140–6; local government in, xix, xxv, 8, 17, 53, 77, 79–80, 84, 91, 159, 161–3, 170; location of, xv, xvi, 43–5, 167; move to, 101–2, 108–11, 125, 168; pedestrian system in, 58, 70, 121, 176; photographs of, xxvi, 68–9, 72–3, 75–7, 114–5, 120–1, 128–9; population of, xv, 46, 52–4, 66, 101, 103–5; problems in, 136–40; rates, 53, 78, 81, 85, 161; schools, 126, 138–9; shopping, 126, 136, 138–9, 156; social characteristics of, 97, 99, 101–8, 136, 156; strikes in, 98–100, 132, 143, 136, 138, 168; tourism in, xxi, 46, 52, 77, 90–1, 137–8, 140, 150, 156, 161; town centre of, 47, 57, 74–6; town manager of, 80–3; town planning of, 46, 50–2, 55, 59–60, 62–3, 65–6, 70, 88, 111, 125, 163–5; water supply, 65–6; women in, 106–7, 110, 123, 129, 133, 136, 149, 167–8
Jabiru Area Working Group, 88
Jabiru Community Survey 1984, background to, xix, 7, 97–101; response rates, 99, 175–77; sample design, 99, 101, 135, 173–8; summary results of, 133–4, 156–7, 165–9
Jabiru East, 37–9, 68, 71, 90, 98, 113, 115, 117–8, 120, 174, 176
Jabiru Rag, 81–3, 155, 165, 175
Jabiru research project, xvii, 99
Jabiru Town Advisory Council, 78–9, 81–4, 88, 90, 150–7
Jabiru Town Council, 78–9, 85, 140, 150–7, 175
Jabiru Town Development Act, 1979, xvii, 56, 59–60, 64, 77–9
Jabiru Town Development Authority, Aboriginal housing, provision by, 73–4, 149; assistance of, xix, 174–5; chairman of, 25, 59, 77, 79, 83, 89, 160, 164; choice of, xvii, 54, 59–60,

159–61; debts of, 66, 85, 160–1; inaugural meeting of, 60–1; Jabiru, construction by, 58, 65, 135; membership of, 59, 60–1
Jacobs, J., xxiii, 171
Japanese, 19, 37
Jim Jim Creek, 52, 140
Jim Jim Falls, 52, 142–3
Jones, R., xxii–xxiii, 33

Kakadu National Park, Aboriginal involvement in, xxii–xxiii, 51, 87, 92, 143, 150; Jabiru residents and, 110, 132–3, 153, 140–6, 164; location of, xv, 52; management of, 73, 141; origins of, xxi, 7, 33, 36; plan of management for, xxiii, 51, 71, 89, 92, 141, 161; visitor usage of, xxi, xxiii, 171
Kambalda, WA, 24
Karratha, WA, 7
Katherine, NT, 8, 55, 161
Keen, I., 171
Keesing, N., 170
Kelly, J.P., xxiii, 171
Kennedy, D. jnr, 170
Kennedy, B.E., 17–18
Kerr, C., 4
Kilmartin, L., 108
Koongarra, xv, 37, 42, 45, 84, 86
Kriegler, R., 27, 134

Labor government, 54, 84, 87, 89, 136
Lake Havasu City, Arizona, 131
Land Act (Western Australia), 1893, 24
Langrod, G., 24
Lansing, J.B., 108
Laver, P., 26
Lawrie, D., 60
Larcombe, J.R., 60
Layman, L., 19, 21
Lea, J.P., xii, xvii, xviii, xx, xxiii, 4, 28, 43, 51, 65, 81–2, 91, 97, 162, 172
Leichhardt, L., 71, 171
length of residence, 101–2, 108, 111, 127, 130, 151, 157
Levitus, R., 171
local government, councillors and aldermen, 16–18, 27, 82, 152–3, 157; Jabiru, introduction of, xxv, 8, 77, 79–80, 84–5, 91, 159, 161–3; nineteenth century in, 15–18; Northern Territory in, 8, 77, 86, 159, 161–2; Pilbara, 18–28; public

meetings about, 16–17, 22–3, 79, 81–2, 165; rates, 26, 53, 78, 81, 85, 161; taxing of, 16–17, 27; Victoria in, 15–17, 29, 77, 163; Western Australia in, 23
Local Government Act (Western Australia) 1960, 26
Local Government Amendment Bill (NT) 1981, 161
local politics, 29, 78
Lodge Agreement, 37, 47
longitudinal analysis, 167
Loveday, P., xii
Lucas, R.A., 4

McHenry, R., 79
McIntosh, A.H., xii, 41–2, 61, 79
Macknight, C.C., 34
Macassan fishermen, 34
Magela Creek, 40, 71
Main, E.M., xii, 79, 82, 112
malaria, 58
Manaburduma camp, 72, 74, 129, 135, 175
Marble Bar, WA, 26
marriage, 123–4, 133, 135–6, 139, 168
Marxist, 4, 6, 19, 87, 159
Maryborough, Victoria, 16
Maryland, USA, 132
Mary Kathleen, Queensland, 108, 111
Melbourne, Victoria, 4, 16–17, 110
Melville Island, NT, 34
Merrill, P.J., xxiii, 171
mining town development, 4–5, 156, 159, 163–5
Minister of the Uniting Church *see* Main, E.M.
Moranbah, Queensland, 6, 134, 137
mosquitoes, 58, 83
Mount Brockman, xxiii, 41, 48, 57, 71
Mt Isa, Queensland, 101
Mount Newman Mining Company Ltd, 22, 24–6, 111
MSJ Keys Young Planners Ltd, 26
Mudginberri, 34, 36, 49, 52, 56, 71, 149
Municipal Institutions Establishment (Victoria) Act 1854, 15–16
Munmarlary, 36, 49
Murdoch University, WA, 22

NABALCO, 78, 92
Nabarlek, 37, 42, 167
Nanambu Creek, 66
nationwide demountable units, 113, 115

National Parks and Wildlife Conservation Act 1975, xvii, 37, 51, 61

Neil, C.C., 103, 106, 110, 119, 135, 173

Newman, WA, 13–14, 22–7, 106, 124, 168

Newton, P.W., 6, 70, 119, 127, 135

New Towns, 18–21, 83, 125

Nhulunbuy, NT, 8, 46, 51, 60, 69, 78, 80, 82, 86, 91–2, 102, 112, 159, 161–2, 167

Nieuwenhuysen, J., xxi

Noranda (Australia) Ltd, xviii, 37–8, 49, 65

normalisation, 21–2, 77, 92, 164

Northern Land Council, xviii, 47, 49, 50, 53, 55–6, 58, 71, 87–9, 150, 162

Northern Pastoral Services Ltd, 49

Northern Territory, Department of Health, 83, 90; Law, 88; Mines and Energy, 88; Transport and Works, 60, 88

Northern Territory Housing Commission, 44–5, 73, 89–90

Northern Territory Labor Party, 81

Northern Territory Liquor Commission, 87

Northern Territory Reserves Board, 36

Northern Territory Treasury, 61, 89

Northern Territory University Planning Authority, xii, xiii, xix

Nourlangie Creek, 52, 140

Nourlangie Rock, 52

Obiri Rock, 52

Oenpelli, NT, 34, 52, 71

Oenpelli Council, 49

Oeser, O.A., 119

O'Faircheallaigh, C., 50

Office of the Supervising Scientist for the Alligator Rivers Region, xvii, 88, 107–8, 139, 166

open towns, 59, 91, 162

Optiz 'Cooinda' Enterprises Ltd, 49

outstations, 56

Ovington, J.D., 171

Owen, J., xii

Paget, G., 4, 163–5

Palmer, K., xxiii, 140, 150, 158, 162–3, 169, 171

Palmerston, NT, 8, 161

Pancontinental Mining Ltd, xv, xviii, 49, 51, 65, 88, 90, 160

pandanus palm, 38

Pannawonica, WA, 22

Papua New Guinea, 101

Paraburdoo, WA, 22, 106

Parkin, A., 29

Parsons, D., 50, 55, 171

participant observation, xix, 6

pastoral industry, xxiii, 33–4, 36, 43, 90

Pattle, G., 112

Peace, A., xix

Peat Marwick Mitchell Services, 171

Peko-Wallsend Ltd, 37, 56, 61

Perth, WA, 18, 23–4, 28, 110, 138

Pilbara, xxv, 6, 13, 18–28, 92, 101, 106, 110–11, 124, 127, 137, 148

Pine Creek, NT, 140

pluralist interests, 9, 90–1

Point Samson, WA, 7

political economy, xx, 19, 28, 80, 159, 168

Port Essington, NT, 34

Port Hedland, WA, 23, 110, 127

privacy, 116–18

private property ownership, 10, 14–16, 28–9, 78, 84, 91, 165

project managers, Jabiru see Cameron McNamara-Minenco Joint Venture

Proud, Sir J., 38

Public Health (Victoria) Act 1854, 16

Queensland Mines Ltd, 37–8

Quilp, 71

Rabnett, R.A., 163–5

Radburn town planning, 6

Raffles Bay, NT, 34

Ranger Agreement, 50, 170

Ranger anomalies, xxiii, 36–40

Ranger Uranium Environmental Inquiry, xv, xxi, xxiii, 33, 37, 41, 47, 49–50, 56, 59, 160, 162, 164–5, 170–1

Ranger Uranium Mines Ltd, Aborigines and, 50, 147, 150; construction of, 56; employment by, 104–7, 138; finances of, xxi; Fox Inquiry and, 49; future of, 136, 169–70; Jabiru Town Development Authority and, xviii, 60, 64, 80, 85; location of, xvi, 40, 45; management of, 97–8

Read, J., 22, 26–7

Redwater Disease, 34

regional centre, 33–4, 43–7, 54

remote communities, 3, 6, 10, 12, 18, 25, 28–9, 74, 131–2, 137, 139, 165–6, 173

Resource Communities Environment Unit, CSIRO, 4, 110, 173
response rates, 99, 175-7
Rhodes, R.A.W., 86
Riffel, J.A., 4
Rillstone, N., 79
roads, 43, 47, 56-7, 61, 65, 70-1, 121, 124
Roberts, J., xxiii, 50, 55, 171
Roberts, S., 50, 171
Robinson, I., 119
Roebourne Shire, WA, 7, 21-2, 106, 137, 148
Roper Bar Trading Pty Ltd *see* South Alligator Motor Inn)
royalties *see* uranium mining
Royal National Park, Sydney, 36
Roxby Downs, xxii, 86
Rum Jungle, 6, 36
Russell, C.C., 60
Ryan, G.R., 36, 38, 41, 171

sacred sites, xxii, xxiii, 41-2
Saddler, H., xxi, xxii, xxiii, 36, 43, 171
Salaman, G., 4
Sanders, W., xii
Sandhurst, Victoria, 13, 15-16
sand mining, xxiii
Saunders, P., 9, 86-7, 90
School of Town Planning, University of NSW, 131
scintillometers, 38
self-government, Northern Territory, xvii, 58, 74, 78, 91, 159, 161
Serle, G., 15-16
Shay Gap, WA, 110-11
Sheringham, N., xii
Siegel, A., 4
Simpson, E.J., 43, 56, 58, 60, 172
single persons, 70, 97, 99, 101-6, 108-26, 130-1, 133, 136-8, 144-5, 148, 151-5, 157, 173-4, 176-8
single persons quarters, 56, 70, 111-13, 115-26, 174, 176
Smith, Jack, 34-5
Smith, Terry, 79, 81
social engineering, 165
social impact assessment, 163
Social Impact of Uranium Mining Project, 56, 88, 171
social mobility, xix, 11, 29, 167
South Alligator Motor Inn, 49, 52
South Alligator River, 34-6, 43, 140
South Australia, 18, 27
South Hedland, WA, 6

Spencer, Sir W.B., 171
Stacpoole, H.J., 17
staging of residential development, 65-6
Stanley, O.G., xii, xxiii, 171
Steelworks Indenture Act (South Australia) 1958, 27
Stimson, R.J., 108
Stockbridge, M., 6, 106, 135, 136, 148
Stolz, G.E., xii, 79, 89, 160, 172
St Martin, I., 4
street names, Jabiru, 70-1
strikes, 11, 24, 98-100, 132, 134, 136, 138, 168
Stubbs, I., 23-4
Sullivan, H., 171
Supervising Scientist *see* Office of the Supervising Scientist
Sydney, NSW, 18, 130-1
Szelenyi, I., 27

Tarnagulla, Victoria, 16-17
Tatz, C., xxiii, 171
teenagers, 133, 135, 137-9, 156
Tennant Creek, NT, 8, 102, 161
Tenthy, N., xii
termite hills, 38
Thompson, F., 6
Thompson, H.M., 4, 8, 13, 21-4, 159
Thorns, D., 108
Tom Price, WA, 14, 22
tourism, 34, 43, 90, 92, 135, 140, 150, 156, 161-2, 169-71
town centre *see* Jabiru
townhouses, 56, 70, 111-12, 114, 118
trades unions, 7, 22-3, 60, 98-100, 134, 151-2, 165, 168, 175
Trollope, A., 17
Tumbler Ridge, British Columbia, 163-4
Twin Falls, 52

UDP Falls, 52
United States of America, xx, 131-2
United Uranium NL, 34
upfront payments, 55
Uranium Advisory Council, 88
uranium mining, controversy over, xx-xxii, 137; discoveries and, 21, 33-43, 171; Jabiru dependence on, 169-70; profits from, xxi; royalties from, 50, 90, 92, 146, 148, 158, 170; social impact of, xxiii, 56, 171
Uranium Province, NT, xv, xxi, xxv, 33, 42-3, 78, 80, 86, 90, 92, 158, 160, 162-3, 169

200   *Index*

uranium question, xiv, xx–xxii, 136–7,
  156, 158, 169–70
Urlich Cloher, D., 15
Utah International Inc., 111, 119

Victorian Legislative Council, 15–17
Von Sturmer, J.R., xv, xxiii, 171

Walisser, B., 4
Walker, A., 6
Warburton, C., 34, 171
Warren, R.L., 3
water supply, 38, 65–6
Weaver, S., xxiii, 171
Weber, M., 10
Weems Report, 37
Weipa, Queensland, 69
West Alligator River, 34–5
West Arnhem Land Co-ordinating
  Committee, 88
West Pilbara Shire, 22
White, C., 171

Whitlam Labor government, xxiii, 46
Whyalla Council, 27
Whyalla, South Australia, 27, 134
Wickham, WA, 7, 22
Wild, R.A., xix, 6
Williams, C., 6, 134, 137
Willoughby, NSW, 130
Witt, K., xii, 112
Wittenoom, WA, 21
Wollstonecraft, NSW, 130
Wood, M., 8, 29
Woods, D.T., xii, 50, 60, 171–2
Woodward, A.E., 47
World Heritage List, xiv

Yellow Water, 52
Young, E.A., xii
Young, R.I., 172
Yulara, NT, 85
Yunupingu, G., 55

Zehner, R.B., xii, xvii–xviii, xx, xxiii, 4,
  28, 43, 81–2, 97, 131–2, 172